Journey from the Center of the Sun

Journey from the Center of the Sun

• JACK B. ZIRKER •

PRINCETON UNIVERSITY PRESS • PRINCETON AND OXFORD

Copyright © 2002 by Princeton University Press
Published by Princeton University Press, 41 William Street, Princeton,
New Jersey 08540
In the United Kingdom: Princeton University Press, 3 Market Place,
Woodstock, Oxfordshire OX20 ISY
All Rights Reserved

Library of Congress Cataloging-in-Publication Data
Zirker, Jack B.
Journey from the center of the sun/J.B. Zirker
p. cm.
Includes index.
ISBN 0-691-05781-8 (alk. paper)
1. Sun I. Title
QB521 .Z58 2001
523.7--dc21

British Library Cataloging-in-Publication Data is available

This book has been composed in Janson and Helvetica

Printed on acid-free paper

www.pupress.princeton.edu

Printed in the United States of America

1 3 5 7 9 10 8 6 4 2

CONTENTS

LIST OF FIGURES

PREFACE

e live in a time of breathtaking advances in astronomy. To mention only a few headlines: the universe's expansion seems to be accelerating, most of the known galaxies lie on the surfaces of enormous bubbles, more than 90 percent of all the mass of the universe remains unidentified, and huge hungry black holes lie at the centers of many galaxies.

Cosmology is in its finest hour. Why then bother with one miserable star that happens to lie near us?

The answer, or one answer, is that there is a certain beauty in the details. When seen from the distance of Alpha Centauri, our sun would be just another pinpoint of light in the sky. However, from our vantage point we can see how this jewel of a star is constructed. No theorist sitting in her quiet cell could possibly imagine the variety and splendor of this star's complexity. It has afforded generations of astronomers with wonder and delight. Most recently, as the technology has developed, we can look inside the creature and begin to understand its metabolism.

I wrote this book because I wanted to share some of my pleasure in the subject. But not merely the pretty pictures one sees in TV specials or on the Web. I wanted to try to explain how the sun works, the physical principles that govern its behavior, the many things we have learned since Sputnik, and the long list of things we still don't understand to our satisfaction.

Because my point of view stresses the physics, some parts of the book are more difficult than others. To help the reader, I've shifted the stickier parts to notes in an end section. So it's possible to read the book in two helpings, the first a quick and hopefully enjoyable run, the second a more leisurely tasting. One can read the chapters in any order, but they may make more sense if read in sequence.

If the book reads well, Trevor Lipscombe, Joe Wisnovsky, and Bill Laznovsky at Princeton University Press can take a lot of the credit. I'm grateful to Gene Floersch for the tremendous effort he put into reworking the figures and planning the layout. I want to thank John Cornett, librarian at the Sacramento Peak Observatory, for his unstinting help in tracking obscure references. The experimental teams of SOHO, TRACE, and the ground-based observing networks deserve my gratitude for making so much of their current research easily accessible on the World Wide Web. And to all my colleagues in the trenches I add my thanks and best wishes.

Journey from the Center of the Sun

INTRODUCTION

 n March 13, 1989, the sun ejected ten billion tons of hot hydrogen gas toward the earth. After a trip of three days this huge mass struck the earth's magnetic field. The impact caused huge electrical currents to surge in the power lines of eastern Canada. Within a few seconds, transformers smoked and relays melted. The entire electric power grid of Quebec shut down. Millions waited in the dark for power to be restored.

This powerful event helped people all over the world to realize, perhaps for the first time, that the sun is more than just a bright ball in the sky. We all normally take the sun for granted. We seldom reflect that all life on earth—from microbes to man—depends critically on the steady outflow of energy from our nearest star. The sun warms us, nurtures our crops, powers our weather. It is, literally, the most important astronomical body for mankind, and yet many people know very little about it.

In this book we'll become better acquainted with our nearest star. We'll take an imaginary journey from its center, following its flow of energy out to the orbit of earth. On the way we'll examine how its parts fit together, and how they work. We'll discover that the sun is not just a passive lump of gas, but has a complicated internal metabolism. It churns and roils, spits gas into space, and emits deadly radiation. Every day, even in its quiet moments, it blows a hot wind of electrically charged particles throughout the solar system.

In the course of our travels we'll learn some of the physical principles that the sun obeys. We'll also examine the questions that solar physicists are still trying to answer.

This is a good time to take such a journey. Researchers are currently harvesting observations of unprecedented quality from several satellite observatories, such as the YOHKOH, the SOHO, TRACE, and Spartan missions and from ground-based telescopes such as the GONG and RISE networks. At the same time the latest generation of high-speed computers has enabled a new breed of astronomers to simulate in great detail the sun's inner workings. All this is leading to rapid progress in understanding our sun. It makes for a fascinating story.

Chapter 1

GETTING STARTED

ou might be surprised at the amount of solar research going on at the moment. Surely, after a hundred years, astronomers must know everything we'd want to know or need to know about the sun. After all, it's so bright, so near, so easy to study that there must be nothing new to discover.

Far from it! We'll see that indeed we do know a lot but that several basic questions that have been dangling for decades still lack adequate answers. Or that different answers have been proposed and all are still controversial. Here, for example are some of the "big" questions solar astronomers are grappling with today.

First is the question of what heats the corona. The corona is the faint outer atmosphere of the sun that one can see with the naked eye during a total solar eclipse. For reasons we still don't know, the corona is five hundred times hotter than the surface of the sun. Since heat can't run "uphill," from cooler to hotter places, something other than the heat and light from the surface must raise the temperature of the corona. A half dozen ideas have been offered over the past twenty years to explain this state of affairs, but observations have shown all of them to be either incorrect or unverifiable. So the quest for answers goes on.

A second big question concerns the solar wind. We know that the corona boils steadily off into space as a solar "wind" that flows around the earth at speeds as high as 800 km/s. This gusty wind produces auroras and so-called geomagnetic storms, events in which the earth's magnetic field takes a terrific buffeting. These storms interrupt radio communications, among other nasty effects. Although the solar wind has been studied intensively, especially from space-craft, we still don't know how it's being accelerated to such high speeds. Again, several competing ideas have been proposed and are being followed up.

We've already mentioned the masses of hot gas that the sun shoots off now and then toward the earth, like the one that zapped Quebec. These are not part of the steady solar wind, but they also originate in the corona and are called "coronal mass ejections." We still have rather fuzzy ideas on how such mass is expelled from the sun (another big question) and clearly we would like to be able to predict such events in advance.

Solar flares are violent explosions in the lower corona of the sun. They can sear the earth with deadly ultraviolet light, X-rays, and solar cosmic rays. A medium-sized flare can release the energy equivalent of a billion megatons of TNT in a few minutes. And such a flare can occur once a day somewhere on the sun. The question is, how does the sun do it? More solar astronomers work toward understanding how energy is stored and released in flares than on any other topic and yet only lately have they gained any real observational support for their ideas.

And then there are questions about the origin of the solar cycle. Everyone has heard of sunspots, the dark patches that cover the sun's face like acne. They are places where powerful magnetic fields (a thousand times the earth's field strength) break through the solar surface. We know the sun is a magnetic star, capable of generating such strong fields somehow in its interior. But exactly how does it do it? And why do the spots come and go in a cycle of about eleven years? These questions are not purely academic, as we shall see, since there is some persuasive evidence that the earth's atmosphere and its climate responds to this cycle.

Finally there's the question of where all the neutrinos have gone. (Recall that a neutrino is a well-known elementary particle with virtually no mass and no electric charge.) The sun generates its energy in a thermonuclear furnace at its center, using reactions similar to those in the hydrogen bomb. In the chain of reactions, a predictable flux of neutrinos is produced. As we shall see, only half of the predicted amount is actually observed. Does this mean that the whole theory of energy generation in stars (which depends critically on the sun) is wrong? Perhaps not. Perhaps the sun is telling us that our understanding of elementary particle physics is incomplete. If so, the so-called Standard Model of nuclear physics, painfully constructed over two decades, would have to be modified. And then an answer might be found to one of today's central cosmological problems, namely, where is all the missing mass in the universe?

These open questions are only a sampler of the subjects solar astronomers and their allies are investigating now. But in addition, new discoveries about the sun are constantly being made.

As an example, we might mention how astronomers reacted to the sun's assault on Quebec. This event, more than others, convinced solar astronomers that they needed to understand how the sun can fling such huge masses toward the earth. At the very least they wanted to find a characteristic warning signal on the sun that such an event is about to occur. But how to proceed? What could they observe?

These mass ejections have a long history. They were first observed in visible light with a special telescope (a "coronagraph") aboard Skylab, a solar satellite that was launched by the U.S. in 1973. A coronagraph produces an artificial solar eclipse, allowing viewers to see the sun's faint atmosphere (the "corona"), outside the solar disk. During an event (see Figure 1.1), one could see part of the corona detach itself and sail off into space. Such events occurred about once a day.

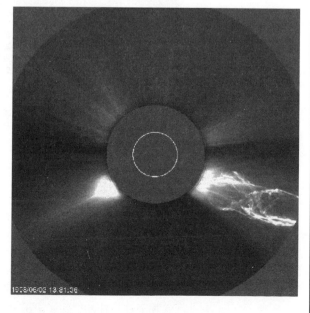

Figure 1.1 A coronal mass ejection in progress. The white circle shows the size of the sun. Also see color supplement. (Courtesy of SOHO/LASCO consortium. SOHO is a project of international cooperation between ESA and NASA).

The Solar Maximum Mission, launched in 1980, showed events more frequently during the peak of the solar cycle, and enabled scientists to study the structure and motions of ejections. But neither of these solar observatories was equipped to find the sources of the ejected mass on the surface of the sun. The reason is that the corona is too faint in visible light to be seen against the glare of the solar disk.

Astronomers had to wait for the launch of the Japanese satellite YOHKOH, in August 1991, to make further progress. This solar observatory was equipped with a sensitive X-ray camera. The corona, as we shall see, is so hot that it radiates X-rays, while the cooler surface of the sun does not. So an X-ray picture of the sun shows the hot bright corona against a dark solar disk. In YOHKOH's pictures (Figure 1.2) the corona is brightest over "active" regions, areas that surround sunspots. In these regions, bundles of bright coronal loops connect different sunspots.

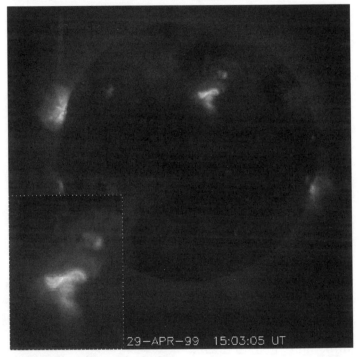

Figure 1.2 The sun in X-ray light. A single active region on the dark disk has an S-shaped coronal magnetic field (shown enlarged in the inset) which is an indication that it may eject coronal mass within a few days. Also see color supplement. (Courtesy of the YOHKOH team. YOKOH is a project of international cooperation between ISIS and NASA).

29-APR-99 15:03:05 UT

In most active regions, the bundles have no special shape. But David Rust, a solar physicist at Johns Hopkins University, noticed that some bundles in the YOHKOH images have a "sigmoid" or S-shape. Then in March 1999, a group of solar physicists at Montana State University announced at a NASA press conference that they had found a long-sought alarm signal that a coronal mass ejection was about to occur. Or more exactly, they found that a majority of ejections were associated with sigmoid bundles of coronal X-ray loops in YOHKOH images. To extract this result they had to study two years' worth of images, make digital movies of them, and correlate mass ejections with sigmoids.

While this result is a distinct advance, these scientists are not yet able to predict exactly when a mass ejection will occur, only that one is likely within a few days after the sigmoid forms.

This is a good example of the new studies, some with practical applications, which are coming out of the latest satellite observations. But what causes the ejections? We will have to defer that question until later in our journey, when we have a better understanding of the corona and its eccentricities.

An Overview

Before we plunge into the center of the sun, we need to view it from a distance, as a star, and to become better acquainted with some of its simplest characteristics. Although the sun is unique to mankind, its type is very common in our galaxy. There are literally billions like it, and many of these probably have planetary systems much like our own. The sun is pretty unremarkable as stars go. Some stars are a hundred times bigger, some a tenth as large. Some are ten thousand times brighter. Some contain ten times the sun's mass, others only a tenth as much. In almost all respects, the sun is a middling kind of star.

Astronomers believe that the sun was born in a huge interstellar cloud of molecular hydrogen gas, some 5 billion years ago. The cloud slowly collapsed under its own gravity, spinning faster and faster as it shrank, much as a skater spins when she folds her arms. Gradually, a flat disk developed at the edges of the cloud, from which all the planets eventually formed. At the center of the cloud, the proto-sun continued to shrink, heating up as gravity compressed the gas. After several million years, thermonuclear processes ignited in the sun's core and it began to shine as a real star. Theorists calculate that it has at least another 5 billion years left before it runs out of fuel. It has settled into a relatively staid middle age, but as we shall see, it also has its violent side.

Most people have trouble grasping how large and massive a star is. (Table 1.1 summarizes the dimensions and other properties of the sun.) If the earth were the size of a period on this page, the sun would be the size of a golf ball. It has a million times the mass of the earth and if that doesn't impress you, consider that the sun loses a million tons a second in the form of sunlight and has been doing that, with no perceptible shrinkage, for over four billion years!

Sun

Earth

Table 1.1 Vital Statistics of the Sun and the Earth		
	Sun	**Earth**
Age	4.6 billion years	4.5 billion years
Average Distance	1.4×10^8 km	----------
Diameter	1.4×10^6 km	1.2×10^4 km
Mass	2.0×10^{30} kg	6.0×10^{24} kg
Average Density	1.4 g/cm³	5.5 g/cm³
Energy Output	3.9×10^{23} kW	----------
Escape Velocity at Surface	618 km/s	11.15 km/s
Central Temperature	15×10^6 K	6000 K??

By human standards, the sun radiates a staggering quantity of energy. A solar hot water heater on your roof could collect over a kilowatt of energy per square meter of area. That number (corrected for the distance to the sun) implies that the sun generates 4×10^{23} kilowatts. In one second the sun emits enough energy to supply the United States for *four million years*!

Most of the energy in sunshine is contained in the visible part of the sun's spectrum, and it is this part we are most familiar with. But the sun also emits X-rays, ultraviolet light, "heat waves," and radio waves. (Figure 1.3 shows the complete spectrum of sunlight.) In general these different wavelengths of light originate in different parts of the solar atmosphere. X-rays, for example, can only escape from the tenuous corona.

Fig. 1.3 The electromagnetic spectrum, which extends from short wavelengths (X-rays) to long wavelengths (radio waves).

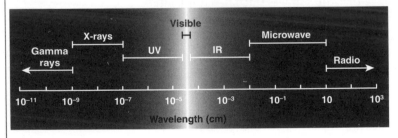

Fortunately for us, the visible light we receive hardly varies in intensity, but the shortest wavelengths (X-rays) and the longest (radio waves) can vary by a factor of ten or more when an explosion (a "flare") occurs on the sun's surface.

Figure 1.4 A sample of Fraunhofer absorption lines in the sun's visible spectrum. Each element emits unique patterns of lines that enable astronomers to determine its abundance in the sun. Also see color supplement.

Practically everything we know about the sun has been learned by analyzing the light it sends us. Figure 1.4 shows a short segment of the sun's visible spectrum. It is riddled with thousands of dark "spectrum lines" that are the fingerprints of different kinds of atoms. By analyzing these lines, solar physicists have determined the chemical composition of the sun (see Table 1.2). Hydrogen is by far the most abundant element, accounting for 70% of the mass. Helium, an element that was originally discovered on the sun, is next in line at 28%, and all the rest appear in tiny traces.

Table 1.2 **The Most Abundant Elements in the Sun**		
Element	Relative Abundance (logarithms)	Relative Mass (%)
Hydrogen	12.00	70
Helium	11.00	28
Carbon	8.7	.41
Nitrogen	8.0	.10
Oxygen	8.9	.91
Neon	8.0	.14
Magnesium	7.6	.06
Silicon	7.6	.06
Sulphur	7.2	.04

The sun spins about a fixed axis, but not rigidly, as a bowling ball would. Because it isn't solid, its different layers can rotate at different speeds. At its visible surface, the sun's equator rotates in twenty-seven days (from east to west, as seen from earth) and in thirty-five days near the poles. As we shall see, the solar interior rotates in a very complicated way that astronomers do not understand as yet.

Normally we can only see the bright surface of the sun, but during a total eclipse (or from space) we can see that it has a huge faint atmosphere, the corona (Figure 1.5). We now know that gas from the corona blows off, continuously and in all directions, as a "solar wind" that streams throughout the solar system at speeds as high as 800 km/s.

With these basic characteristics in mind, it's time for a closer look at what we're about to see on our journey.

A ROAD MAP

We can think of the solar interior as built up of spherical layers, and our first task before starting out, is to become familiar with them. Figure 1.6

Figure 1.5 The visible corona at the solar eclipse of July 11, 1991. Streamers are at their most beautiful near the maximum of the solar cycle. (Courtesy of the High Altitude Observatory. Also see color supplement.

should help. It shows a cutaway view of the sun, revealing its interior and some of its external appendages.

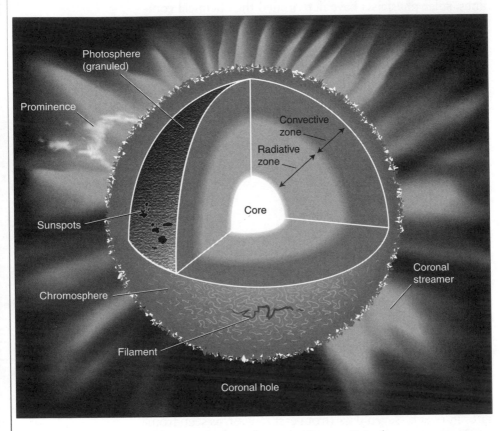

Fig. 1.6 A cutaway view of the sun, with labels identifying various parts of the sun. (Courtesy of the GONG project team, National Solar Observatory)

The core is the solar powerhouse. It is there, at a temperature of 15 million kelvin, that the same process that powers the hydrogen bomb generates the energy that finally escapes as sunlight. In the process, hydrogen is converted to helium. Later on we'll follow the steps in this process, learn why the sun doesn't explode like a bomb, and ask where all the missing neutrinos have gone.

Above the core lies the so-called radiative zone, which occupies most of the solar interior. This region is as opaque as a brick wall and yet sunlight manages to trickle toward the surface. We'll see how this happens later.

At a point about three-quarters of the solar radius from the center, the sun's gas begins to churn at speeds of a few meters per second. Bubbles of hot gas, the size of Alaska, rise to the surface, radiate their heat, and sink back to the depths. This region is the convection zone. It would be easier to understand how convection works in the sun if it weren't so turbulent, but turbulence is still one of the great, unsolved problems in classical physics. Later on, we'll see how computer simulations are helping to connect turbulence to the trendy subject of "chaos."

At the top of the convection zone lies a thin layer, the photosphere, where sunlight finally escapes into space. As a ball of gas, the sun has no firm surface, but the photosphere is the nearest thing to one. It's a sharp boundary between the relatively dense interior and a tenuous corona that extends many solar radii into space.

It's ironic that, at the moment, the sun's interior seems easier to understand than its outer atmosphere, the part we can see with our eyes. One reason for this odd state of affairs is the development of a new observational tool, helioseismology. The surface of the sun vibrates, like the top of a drum. In recent years, astronomers have learned to interpret the sun's tiny pulsations and to derive from them detailed information on the solar interior in the same way seismologists probe the earth's interior. It's as though they can look straight to the center of a star. On our journey we'll examine what they have discovered.

Maybe the sun's interior is actually simpler than the atmosphere or maybe we just haven't resolved its detailed structure yet. What is certain is that the complicated details we can now see in the atmosphere are challenging the best of solar astronomers.

As a useful rule of thumb, the higher you look in the solar atmosphere, the hotter and more rarified it becomes. In the "chromosphere," say, 2,000 km above the photosphere, the temperature is about 10,000 kelvin, while in the corona, at a height of 20,000 km, the temperature is over a million degrees. Why the atmosphere's temperature rises in this way is one of the most challenging questions astronomers are trying to answer. One thing is sure, the sun's magnetic fields are crucial.

The strongest magnetic fields are found in sunspots, the dark, relatively cool patches on the sun's face (Figure 1.7). How the sun creates sunspots and why their numbers oscillate in cycles of 11 or 22 or 88 years remains uncertain, but we'll review what the experts know. Incidentally, you could fit the earth comfortably inside the dark center (the "umbra") of the small spot on the lower left in Figure 1.7.

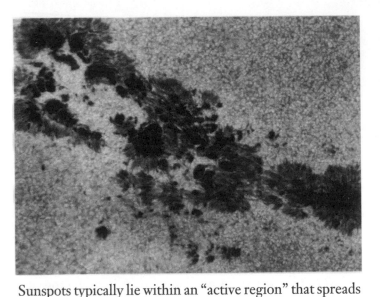

Figure 1.7 A huge cluster of sunspots as seen in visible light. Strong magnetic fields emerge from the dark "umbras" of the spots. The speckles in the picture are "granules," convective cells each about the size of Texas. (Courtesy National Solar Observatory)

Sunspots typically lie within an "active region" that spreads around and above them as a tangle of hot loops. Figure 1.8 shows a typical active region in soft X-rays (with waves of 1 to 10 nanometers). The short low loops have temperatures up to about half a million kelvin, while the higher faint loops have coronal temperatures. All this loopy structure owes its existence to magnetic fields that sprout from the surface and rise up into the atmosphere.

Figure 1.8 An X-ray picture of a solar flare in progress in an active region on the disk of the sun. (Courtesy the YOHKOH team)

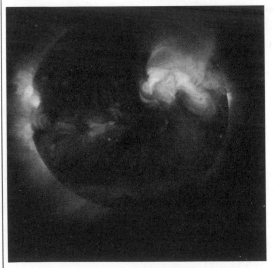

In Figure 1.8 we see this same region in the act of exploding in a solar flare. Its X-ray brightness has suddenly increased by a factor of ten or more. A big flare like this releases as much energy as the entire sun does in a hundredth of a second, mainly in the form of ultraviolet light, X-rays, and the "shrapnel" that it ejects. Solar physicists are hard at work trying to understand how all this energy is stored and released so quickly. We'll find out soon what answers the experts are offering.

Twenty years ago, astronomers thought the magnetic fields in active regions were fairly uniform, at least outside the sunspots. Imagine their surprise when they learned that in fact the fields emerge from isolated patches too small to resolve with even the best telescopes. This arrangement holds even far from active regions in the "quiet" sun. How these tiny patches form and persist is a question we'll consider on our journey.

The coronal streamers one sees during a total solar eclipse (Figure 1.5) are actually long flat sheets of gas that extend many radii from the sun, like the petals of a flower. Their tapered shapes are determined by their magnetic fields and the gas flows inside and around them. Because their temperatures are over a million kelvin, they glow in X-rays against the cooler, darker photosphere. As we'll see, the latest observations from SOHO are helping astronomers to probe such streamers.

Finally, the hot solar wind boils off the sun at a rate of a billion tons a second. When it reaches the earth it contains a mere five protons per cubic centimeter, but still has the power to bend the earth's magnetic field.

With all this as background, we're ready to start our journey.

Chapter 2

THE SECRET HEART OF THE SUN

o begin our journey, let's imagine that we're at the center of the sun, comfortably housed in a fully equipped physics laboratory with perfectly insulating walls. Like explorers in a deep-sea submarine, we can look outside using our remote cameras, perform experiments on our surroundings, and try to interpret what we see.

A glance at our instruments tells us the temperature outside is a blistering 15 million Kelvin—hot enough not only to vaporize all the elements, but also to strip their atoms of most of their electrons. When we zoom in on the gas outside, we find it is a "plasma," composed of free protons, electrons, neutrons, and helium nuclei. Although the material here is 160 times denser than water or ten times denser than the center of our earth, these elementary particles are tearing around at speeds of hundreds of kilometers per second. Like popcorn on a griddle they bounce around furiously and collide many times per second.

As we continue to watch, two protons approach each other at blinding speed in a head-on collision. There is a flash of light and the two protons merge to form a compound nucleus. In addition, two new particles suddenly appear out of the collision and disappear at nearly the speed of light. When we rewind our tapes and replay this event, we find that the total energy of the three particles coming *out* of the proton collision was greater than that of the two protons going *into* the collision.

What happened here? Was the law of conservation of energy violated? Not really. What we saw was the *conversion* of some of the mass of the protons into the kinetic energy (i.e., energy of motion) of the three final particles. As Einstein taught us, mass and energy are equivalent, ($E = mc^2$), and are convertible from one into the other.

So the protons each lost a bit of mass, which showed up in

the kinetic energy of the three particles coming out of the collision. These fast particles immediately collided with their slower neighbors in the plasma and shared their excess energy. In effect the plasma was *heated* by a conversion of mass to energy.

We have been privileged to watch the sun feeding on its own mass to generate the energy that ultimately escapes as sunlight. The idea that such a process could occur took over a hundred years to germinate in the minds of scientists. It's worth going back for a brief look at how this happened.

A Bit of History

What keeps the sun shining? Nobody worried too much about this simple question until a few scientists estimated how much energy the sun radiates. Then it became a difficult problem in physics. It all started with the distance of the sun to the earth.

In 1672 the size and distance of the sun were finally determined with some accuracy by Giovanni Cassini, director of the Paris Observatory, and by John Flamsteed, the first Astronomer Royal. When they took up the problem, the relative distances of the planets to the sun were known from their orbital periods, but not their absolute distances, in kilometers. Cassini's basic method was to triangulate Mars, using the earth's orbital diameter as the baseline of long triangles and measuring the directions of Mars at different places along the earth's orbit. Cassini found a distance to the sun of 22,000 earth radii, fairly close to the modern value of 23,454.8 earth radii, or 149 million kilometers.

Isaac Newton, alchemist, astrologer, and one of the three greatest physicists of all time, was first to measure the sun's radiation, early in the eighteenth century. John Herschel, the son of the great English astronomer William Herschel, improved on Newton. In 1837 he exposed a bowl of water to sunshine and measured its rise in temperature during a fixed period. The French physicist Claude Pouillet obtained an approximate measurement of the "solar constant" by much the same method at about the same time.

When extrapolated back to the solar surface, the amount of energy emitted was staggering, the equivalent of tens of thousands of kilowatts per square meter. Moreover, the sun had presumably been shining at this rate throughout human history—thousands of years. What process could possibly account for this incredible output? Certainly, no chemical reaction, like the burning of coal, could supply such power.

Two explanations were soon proposed. John Waterston, a Scottish engineer, showed that gravitational contraction of the sun at a rate of a hundred meters a year could supply its heat. As the sun shrank, its interior would be compressed and therefore heated. Alternatively, Julius Mayer, a German physicist, proposed that a vast number of meteors, left over from the formation of the solar system, might heat the sun as they fell into it.

Two eminent scientists picked up the ideas of these relatively obscure authors in the nineteenth century. Hermann von Helmholtz, a German scientist with extraordinarily broad interests, developed the contraction theme. The idea that the sun was contracting slowly fitted in nicely with the nebular hypothesis of the German philosopher Immanuel Kant. Kant had proposed that the solar system originated in the contraction of a great cloud or nebula of cold interstellar gas. So a residual contraction of the sun seemed entirely natural to Helmholtz. He estimated that a contraction of a mere 35 kilometers every thousand years could supply the sun's energy loss.

William Thomson, later Lord Kelvin, adopted Mayer's meteor hypothesis. Kelvin was one of the founders of the science of thermodynamics and ultimately became the president of the Royal Society of London. His considerable prestige helped to establish the meteor idea. With a little more thought, however, Kelvin realized that a meteor infall sufficient to heat the sun would also heat the earth's atmosphere to uncomfortable levels. Moreover, the sun would gain mass so rapidly by this process that the planets would be accelerating, contrary to observations.

So Kelvin made a diplomatic retreat. He wrote that although meteoric heating had probably been important in the early history of the sun, it must have ceased. Following Helmholtz, he then adopted the contraction hypothesis. He estimated that the sun, by contracting steadily, could shine at its present rate for some 20 million years.

But by the mid-nineteenth century such a time scale was no longer sufficient. Geologists such as Charles Lyell were uncovering evidence from the present rates of sedimentation and the thickness of sedimentary layers that the earth and the sun were at least several hundred million years old. Nevertheless, Kelvin stubbornly argued against the possibility of such enormous ages and continued to support his contraction hypothesis. His huge reputation was sufficient to keep the contraction idea alive for the next fifty years.

These fifty years saw tremendous advances in physics, chemistry, and astronomy. A century earlier, Isaac Newton had used a glass prism to show that white sunlight contains all the rainbow colors, spread out from blue to red, in the solar spectrum. With better prisms, Josef Fraunhofer discovered that the spectrum was sprinkled with thousands of spectral "lines," wavelengths where the intensity of light was somewhat reduced (see Figure 1.4).

These "absorption" lines were later analyzed by Gustav Kirchhoff and Robert Bunsen. Two strong lines in the yellow part of the solar spectrum were known to arise from sodium atoms. In a crucial experiment, Kirchhoff passed sunlight through a heated sodium vapor. To his surprise he saw the solar lines darken appreciably. He realized that any substance that emits light (the sodium vapor in this case) also absorbs light of the same wavelength. Strong emitters are also strong absorbers. This result led Kirchhoff to conceive of a "black body" that absorbs all incident radiation and must emit radiation of a unique kind. This laid the basis for Max Planck's later work.

Kirchhoff continued his spectroscopic studies. By comparing the wavelengths of Fraunhofer's lines with laboratory sources, Kirchhoff began to identify terrestrial elements in the sun, such as calcium and iron. He launched the science of spectral analysis, paving the way to the complete chemical analysis of the sun and stars. All the terrestrial elements were eventually found in the sun (see Table 1.2). In a curious reversal, helium, the second most abundant element, was first discovered in the solar spectrum by Norman Lockyer.

During these critical years of the nineteenth century, Sadi Carnot, William Thomson, Rudolph Clausius, and J. Willard Gibbs discovered the laws of thermodynamics, which govern the conversion of different forms of energy. Ernest Rutherford's discovery that most of the mass of an atom is contained in a tiny positive nucleus led to Niels Bohr's model of the hydrogen atom and later to the rise of quantum mechanics. By 1905, Albert Einstein's special theory of relativity had revealed the equivalence of mass and energy, that is, that they are convertible from one to the other by the equation $E = mc^2$. Then a critical experiment by Rutherford and Frederick Soddy showed that the mass of a nucleus was smaller than the mass of an equivalent number of protons. That meant that an amount of energy (the "binding energy"), equivalent to the missing mass, would have to be supplied to separate the different particles inside a nucleus.

Arthur Eddington, the distinguished British astrophysicist, made the next great step forward around 1920. He had completed a wide-ranging study of the energy and pressure equilibrium of stars and had constructed mathematical models of temperature and density inside stars. For the sun, he estimated a central temperature of some 40 million degrees and a density of 80 grams per cc. When a critic scoffed at such a high temperature, Eddington told him that if he could find a hotter place anywhere, he should go there!

In a major discovery, Eddington found that a simple relation must hold in general between the total rate of energy loss from a star (its "luminosity") and its mass. Knowing the sun's mass he could predict its luminosity within a factor of two, a wonderful result.

The need for a source of stellar energy lasting billions of years was generally recognized by now. Eddington offered two suggestions. The first was that electrons and protons, having opposite electric charge, annihilate each other in the stellar interior, with a conversion of mass to energy. His second idea, which ultimately won out, was the construction of heavy atoms by the fusion of protons, with some conversion of mass to energy.

When Eddington offered these ideas, physicists had only the most primitive understanding of nuclear processes and their rates. Even worse, although many terrestrial elements had been identified in the solar spectrum, no method was available to establish their relative abundances in the sun. Thus the sun's chemical composition was generally assumed to resemble the earth's, with a predominance of heavy elements like iron.

This misconception was cleared up in 1925 by the work of one of the few female astronomers of the time, Cecelia Payne. She was a tall, fair-haired lady, speaking with great poise in a soft, English-accented voice. In her doctoral thesis, she used a new theory of the ionization of elements, due to Mahindra Saha and Ralph H. Fowler, to analyze stellar spectra. Her amazing conclusion was that the sun and stars are composed almost entirely of gaseous hydrogen.

Henry Norris Russell, the renowned Princeton astronomer, scoffed at the idea. In the culture of the times, a mere graduate student, no less a female, was expected to defer to her male seniors. But Payne held her ground. So Russell decided to see for himself. He reanalyzed the absorption spectrum of the sun, using the new ionization theory, and found to his surprise and delight that Payne was correct. Payne went on to become a world authority on the spectra of stars and a full professor at Harvard University.

This discovery set the stage for the final breakthrough. In 1928, George Gamow showed theoretically how a positively charged nucleus of helium (an "alpha particle") could exit a radioactive nucleus of uranium despite the electrical forces of attraction. In effect the alpha particle "tunneled" its way out of the atom. Two of Gamow's colleagues realized that the process could be reversed; that an alpha particle could tunnel its way *into* an atom, thus creating a heavier nucleus. This demonstration attracted the best nuclear physicists of the time to the critical problem of stellar energy. Such luminaries as J. Robert Oppenheimer, Edward Teller, Lev Landau, and Carl von Weizsäcker all tried to solve the solar energy problem by building heavy nuclei from lighter ones.

Finally, in 1936, Hans Bethe and Charles Critchfield independently conceived the "proton-proton" chain of nuclear reactions that would fuse four protons into a nucleus of helium. Bethe and von Weizsäcker soon found another chain, the carbon-nitrogen chain, which could be effective in stars more massive and hotter than the sun.

TUNNELING

After this brief excursion into history, let's return to our laboratory at the center of the sun, to study these nuclear processes in greater detail.

Each particle, we find, regardless of its mass, has an average kinetic energy fixed by the temperature, T, and by Boltzmann's universal constant, k. An electron, for example, has an average energy kT equal to 1.4 kilovolts. (A kilovolt is the energy the electron would gain passing through a battery with a thousand volts between its terminals.) Protons and alpha particles, despite their larger masses, have this same average energy because it depends only on the temperature.

Let's look once again at two protons colliding head-on. Since each of them has the same unit positive charge, a force of electric repulsion tends to slow them down as they approach each other. To advance closer they must lose some of their kinetic energy and the amount they must lose rises very steeply the closer they get (see Figure 2.1). We call the energy required to bring the protons together, against the force of repulsion, their *potential* energy.

It is in effect a barrier, a "hill" the protons must "climb" to reach each other. Like a ball rolling up a slope, the more

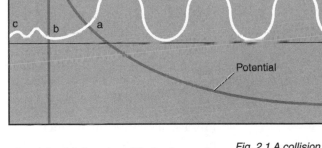

kinetic energy a proton has, the higher it will climb up the potential hill.

This potential hill has a top, because if the protons come close enough (within a proton radius) the so-called "strong" nuclear forces of attraction will overcome the electrostatic force of repulsion. One proton will lose its electric charge to become a neutron, and it then joins with the other proton to form a new particle, a deuteron.

There is a problem however. The height of the hill in this case is about two *thousand* kilovolts. Thus, with only 1.4 kilovolts of energy, the average proton hasn't nearly enough energy to get over the potential hill and join its colliding partner. Of course some protons have more energy than the average and others less. Their numbers are governed by the Maxwell–Boltzmann law, which we will encounter later on. Suffice it to say that at a temperature of 15 million kelvin the number of protons with 2,000 kilovolts of energy is negligible. If protons were just like cannon balls, ruled by the laws of classical mechanics, they would stop

Fig. 2.1 A collision of two protons. One proton is at rest at the extreme left edge of the diagram, the other approaches from the right. The energy required to bring them together against their mutual repulsion is shown as the "potential energy" curve. The instantaneous position of the approaching proton is determined by its accompanying probability wave. At point a this wave tunnels through the potential energy barrier b to reach its target c.

rolling far short of the top of the hill and the story would end here—protons could never merge and the sun could never shine.

And yet we can watch deuterons forming here at the center of the sun.

The escape from this dilemma is the fact that protons are governed by quantum mechanics, not classical mechanics. In quantum mechanics, elementary particles like protons are described not as hard spheres but as fuzzy clouds. To locate a particle one has to resort to its associated "probability wave," which is governed by Schrödinger's wave equation. The strength of the wave at any point in space is a measure of the probability of finding the physical particle at that point. For example, in free space, far from any other charged particles, the wave of a proton would be sinusoidal, as shown in Figure 2.1, with a wavelength equal to $h/(2\pi mv)$, where m is the proton mass, v its velocity, and h is the universal Planck constant. The proton is most likely to be found at one of the maxima of the wave, but since there are many of these, it could be anywhere.

When two protons approach each other toward a collision (Figure 2.1) the probability wave is sinusoidal at large separations, has a gradually increasing wavelength as it approaches the potential hill, and then switches over to a smooth sharply decreasing curve *inside* the potential hill. This means that there is a small but finite probability that a proton with insufficient energy to top the hill, can nevertheless *tunnel through it and reach its target*. This is a strictly quantum mechanical effect, discovered by George Gamov in 1928.

THE PROTON-PROTON CHAIN

Gamov's discovery opened the way for Hans Bethe and Charles Critchfield to construct a chain of known nuclear reactions that could convert hydrogen to helium in the core of the sun with a net release of energy. The basic reason why energy is released is that the four particles that make up a helium nucleus (two protons and two neutrons) have a slightly larger total mass *before* they combine than *afterward*. Their loss of mass m is equivalent to a loss of energy, E, according to Einstein's famous equation, $E = mc^2$. This same amount of energy E would have to be supplied to tear apart the constituents of a helium nucleus; it represents the *binding energy* of the nucleus.

In endnote 2.1, I show the details of the proton-proton chain for those readers who may want more information. A second chain is also there, the so-called carbon cycle, which is important in stars hotter than the sun.

We should clear up one point before proceeding. The thermonuclear proton-proton chain in the sun is the same as that in the hydrogen bomb. So why doesn't the sun explode?

Thomas G. Cowling, an eminent English theorist, supplied the answer to this question of stability in 1935, well before the bomb had been invented. He pointed out that the rate of nuclear energy generation is extremely sensitive to temperature, varying as the temperature raised to the fourth power, T^4. That means that a tiny increase in the core temperature would result in a huge increase in the energy output and, locally, a large heating effect. This could lead to a further energy release, more heating, and eventually an explosion.

Fortunately the core gases would respond instantly by expanding when heated. The expansion would occur so quickly that no heat could be lost to the overlying layers. In such an "adiabatic" expansion, the core pushes against the overlying layers, lifting and compressing them slightly. This action requires an expenditure of the core's internal heat energy. So the core cools because of the expansion and the rate of energy production, which is highly sensitive to temperature changes, returns to its equilibrium value.

THE STANDARD SOLAR MODEL

Bethe found two thermonuclear chain reactions that could be working in the sun. Perhaps there could be others. How do we really know which chain is the most important? A minimum requirement would be to show that an assumed chain produces the observed energy output of the sun. This was first done by Bethe, who used the nuclear rates for the proton chain that were known at the time. He came within 50% of the correct answer, which seemed very convincing. He also found the carbon chain provides only about 5% of the energy. Since then the rates have been measured or recalculated and the comparison has improved.

Beginning about 1865, astrophysicists have developed and

refined mathematical models of the temperature and density throughout the sun. Since Bethe's discovery, the proton-proton chain has been assumed as the principal source of energy production.

As the power of computers has improved, so have models of the solar interior. The most recent models are evolutionary. They start from the epoch when the sun has just collapsed from an interstellar cloud and follow the slowly changing content of helium in the core, until the present age of the sun, about 4.5 billion years. If a model then predicts the correct luminosity and diameter, it is counted as a success. Maybe even "true."

By comparing models and observations, astrophysicists have reached a consensus on a "standard" model. It is based on the following additional assumptions:

1. The sun was born with a homogeneous initial chemical composition, which can be determined from studies of comets and meteorites.

2. Mass is neither gained nor lost during the sun's evolution.

3. Internal rotation and magnetic fields can be ignored.

4. Matter does not circulate within the sun's core, nor does the manufactured helium escape the core.

The standard model is called that because, employing the most accurate nuclear rates, it predicts the total energy output of the sun (its "luminosity") as well as its radius, at its present age. Its success gives astrophysicists confidence that they understand how the stars shine. With that success in mind, they built up an impressive theory of the evolution of stars, based on thermonuclear reactions.

But this was all just theory. There was no independent experimental evidence for or against the theory. Then, in 1964, Raymond Davis, a nuclear physicist at the Brookhaven National Laboratory, began to design a clever experiment to detect solar neutrinos. As we shall see, his measurements have roiled the world of astrophysics for over thirty years.

NEUTRINOS

Imagine an elementary particle without mass, without an electrical charge, and with no magnetic properties, that moves in vacuum nearly at the speed of light. Such a ghostly creature is the neutrino. It is completely unaffected by electromagnetic forces or by the strong nuclear binding force. As a result neutrinos interact with matter extremely weakly.

If you take a quick look at endnote 2.1 you will see that the proton-proton chain produces two neutrinos for every 27 million electron volts (Mev) of energy. So using the sun's observed luminosity we can estimate the number of neutrinos that should escape the sun each second. Figure 2.2 shows

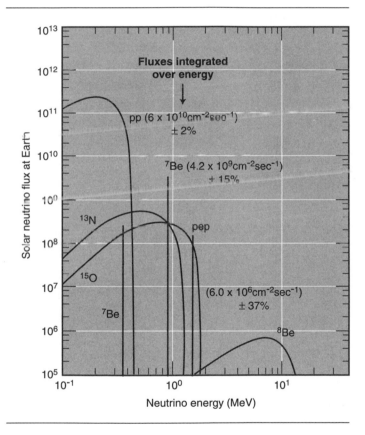

Fig. 2.2 The neutrino fluxes at various stages of the proton-proton chain, as predicted by the Standard Solar Model. (Courtesy of University of Arizona Press)

the predicted numbers of neutrinos with different energies that are emitted by the sun. If we could measure those numbers we could decide whether or not the proton-proton chain really is there.

Nearly all neutrinos produced in the core of the sun travel directly to the earth without ever colliding again. Indeed, most of those arriving at earth pass directly through it as well. In fact, it would take a column of lead a light-year thick (10^{13} km) to stop a neutrino! How then could we possibly measure the number leaving the sun?

Massless neutrinos interact with other matter only through the so-called "weak" nuclear force. This force appears in the decay of radioactive nuclei. For example, a proton can decay into a neutron, with the emission of a positron and a neutrino. A similar weak-force process occurs with many radioactive nuclei. Consider, for example, argon, a noble gas in the same chemical family as helium. Argon has several isotopes, among them a radioactive version with a nuclear mass of 37 protons, ^{37}A. This isotope decays spontaneously into a chlorine nucleus, ^{37}Cl, after a typical "half-life" interval of thirty-five days. An electron and an anti-neutrino are emitted in the decay.

This reaction is reversible if we replace an anti-neutrino with a neutrino. Then a ^{37}Cl nucleus can interact with a neutrino through the weak force to form a radioactive ^{37}A nucleus. *This* reaction, the conversion of chlorine to argon, is the one that Raymond Davis used to detect solar neutrinos.

Incidentally, an anti-neutrino differs from a neutrino, not by an opposite electric charge (as a positron differs from an electron) since neither has such a charge, but by its *spin*. The anti-neutrino has a right-handed spin about the direction of its forward motion, the neutrino a left-handed spin.

CAPTURING NEUTRINOS

Picture a tank 48 feet long and 20 feet in diameter, filled with cleaning fluid, and buried a mile underground in a gold mine (Figure 2.3). Such a contraption hardly fits one's picture of a modern elementary particle physics experiment. Nevertheless that is what Raymond Davis's neutrino detector looks like. It contains 600 tons of perchlorethelene, a volatile compound rich in the chlorine isotope, ^{37}Cl. Compared to a light-year of lead, the tank's power to stop neutrinos is small, but Davis's technique is sensitive enough to detect *individual* captures within the tank (see endnote 2.2).

For nearly thirty years, Davis and his colleagues tended his experiment, carefully counting the rate at which the sun emits neutrinos. If the theory of the proton-proton chain is correct, then every conversion of four protons to a nucleus of helium releases two neutrinos and 26.7 Mev of energy. Since the sun emits 4×10^{23} kilowatts, it should also emit 2×10^{37} neutrinos per second.

This output is truly staggering. As you sit and read this

Fig. 2.3 Ray Davis's
neutrino detector at
the Homestake mine.
(Courtesy of University
of Arizona Press)

book some 10^{10} neutrinos pass every second through every square centimeter of your body! But pass they do, unnoticed. Of course Davis's experiment captures only a minuscule fraction of this vast stream and he has to extrapolate his capture rate to find the rate at which the sun emits neutrinos.

You may ask at this point how Davis's tank is able to capture *any* neutrinos if, as was stated earlier, it takes a light-year of lead to stop one. The answer lies again in probabilities. On average the chance a neutrino will collide with a chlorine atom is almost zero, but not quite zero. If Davis waits long enough, so many neutrinos will enter the tank that this small but finite probability guarantees that one will be captured. If you pick enough clover you will find, sooner or later, a four-leafed one.

Davis reported his results in Solar Neutrino Units, or SNUs. One SNU represents the capture of a neutrino once a second for every 10^{36} chlorine atoms. From the very first re-

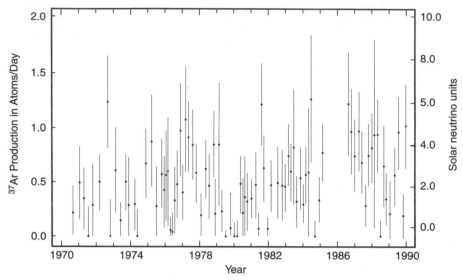

Fig. 2.4 Eighteen years
of neutrino counts by
the chlorine experiment.
The units are in argon
atoms per day.
(Courtesy of University
of Arizona Press)

Fig. 2.5 Neutrino counts
versus sunspot
frequency. The sunspot
curve has been inverted
for a better comparison.
(Courtesy of University
of Arizona Press)

ports in 1970, Davis's measurements have been markedly
smaller than the proton-proton chain theory would predict. His annual measurements (see Figure 2.4) have fluc-
tuated by as much as a factor of ten, even hitting *zero* occasionally, but never quite reaching the best contemporary
theoretical value.

During the interval 1970 to 1990, the average capture rate
was 2 to 3 SNUs. In comparison, the predicted rate, based
on the standard model of the sun, is 7 SNUs *or a factor of
three larger.* Here indeed is a glorious puzzle for astrophysicists! Where are the missing neutrinos?

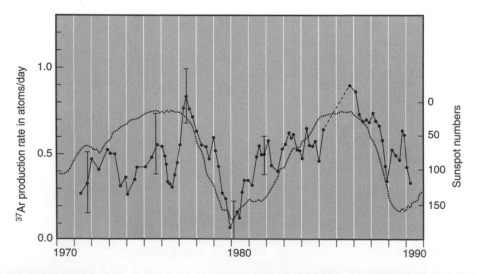

Even more puzzling is the anti-correlation of the numbers of neutrinos and sunspots (see Fig 2.5). This relationship was temporary and disappeared after 1990.

POSSIBLE EXPLANATIONS

Many explanations have been offered for the solar neutrino deficit during the past three decades. Perhaps the most radical (suggested when Davis counted *no* neutrinos) was that the sun had actually stopped producing energy and was slowly cooling down. Eventually the count rate increased, however, and theorists turned to other possibilities.

Some theorists believe these explanations fall into three classes. The first is that the nuclear rates used to calculate the neutrino emission were wrong. The second is that the physics in the Standard Model of the sun is incomplete in some way. And the third is that some crucial nuclear physics is missing.

The first two possibilities have not been completely eliminated but are thought to be much less likely than the third. Nuclear rates have been scrutinized and remeasured to the point where physicists are thoroughly confident in them. A modification of the Standard Solar Model is more difficult to reject.

Any relaxation of the assumptions of the Standard Model produces a nonstandard model, and many have been constructed in order to explain the neutrino deficit. It is great fun and a harmless sport. However, most of these models run into some conflict with other observations or well-established physics. But with creative tinkering, a nonstandard model can reduce the predicted neutrino capture rate to about 5 SNU—a mere factor of two too large.

Some very creative ideas that involve new particle physics have been proposed in the third class of explanations. If, for example, the neutrino behaved like a little bar magnet (i.e., if it had a magnetic "moment") it could flip its spin in a magnetic field and become an anti-neutrino, which Davis's experiment could not detect. As we shall see, the sun's atmosphere does have magnetic fields, but they are so weak that neutrinos would require an unacceptably large magnetic moment to perform their flip.

A very attractive explanation of the third class was proposed in 1978 by L. Wolfenstein and developed further by S. P.

Mikheyev and A. Smirnov in 1986. In essence these authors suggest that during their transit through the dense interior of the sun, neutrinos change into a form that Davis's experiment cannot detect. This proposal resembles the spin flip idea, just mentioned, but the physics is entirely different.

Solar neutrinos are predicted to be solely *electron* neutrinos. We designated such neutrinos in endnote 2.1 with a subscript "e" to emphasize that fact. The conversion of chlorine to argon in Davis's tank also involves such a neutrino. In other weak-force reactions, however, neutrinos are emitted with two other kinds of low-mass particles ("leptons"), namely, muons and tauons and their anti-particles.

Now it turns out that the neutrinos associated with these three types of leptons (electrons, muons, and tauons) are different. They are distinguished in having different "flavors," in the colorful language preferred by particle physicists.

In the strange world of quantum mechanics an electron neutrino is not *absolutely* an electron neutrino until it is measured as such. When a neutrino is emitted in the core of the sun it is a tripart entity—a mixture of the three flavors—partly electron, partly muon, and partly tauon neutrino. As the neutrino travels through the sun the mixture oscillates as a result of weak scattering by solar electrons. The electron flavor interacts with electrons, the other two flavors do not. Thus there is a finite possibility that upon emerging from the sun the neutrino behaves more like a muon neutrino or a tauon neutrino than an electron neutrino. Then when it enters Davis's tank, it will not convert ^{37}Cl to ^{37}A and is lost in his count of solar neutrinos. For a wide range of solar conditions, the flux of emergent electron neutrinos could be reduced by factors as large as four. Et voilá! The neutrino deficit is explained!

For this neat mechanism to work, at least one of the flavors has to have a finite mass, in violation of existing measurements and of the precepts of the standard theory of the weak nuclear force. Estimates of the required mass range from 10^{-6} to 10 electron volts. (For comparison the mass of an electron is 510,000 electron volts.) Such an extension of the standard theory doesn't seem to revolt particle physicists. In fact, an extension seems almost necessary since the present theory doesn't determine many of the physical parameters that enter into it. Moreover, a sufficiently massive neutrino could help to explain another puzzle in astrophysics—the "missing mass problem."

Observations of the motions of galaxies within clusters of galaxies, and of the rotation of individual galaxies, imply that far more mass exists in these objects than their stars and gas clouds contribute. Some form of "dark matter" exists and may account for as much as 90% of all the mass in the universe. Theorists have advanced a variety of possible explanations, including hypothetical WIMP and MACHO particles. But massive neutrinos, left over from the creation of the universe, seem less radical. Unfortunately a neutrino mass of around 30 electron volts would be required to solve the dark matter problem and the present experimental limits seem to rule out this possibility

THE SECOND GENERATION OF NEUTRINO DETECTORS

For two decades, Davis's neutrino experiment was the only one in the world. His unexpected results and their important implications for particle theory and stellar evolution stimulated other groups to set up new experiments, partly to check his results and partly to extend them.

Davis's experiment is sensitive to solar neutrinos with energy greater than 0.814 Mev, which are produced mainly by the decay of the ^{8}Be nuclei. However, as Figure 2.2 shows, the *primary* proton-proton reaction produces neutrinos with energies below 0.420 Mev. In order to sample these low-energy neutrinos two new experiments (GALLEX and SAGE) have been built, employing the rare metal gallium as a detector (see endnote 2.3 for details).

Data from GALLEX were taken from 14 May 1991 to 29 April 1992 and the analysis yielded a solar flux of 81± 17 SNUs. A second series of runs, lasting from 19 August to 3 February 1993 yielded 97 ± 23 SNUs. These two results are consistent with each other within their estimated statistical errors and they correspond to about 65% of the flux predicted by the Standard Solar Model. This deficit is half as large as that of Davis's chlorine experiment, but still worrisome. In comparison, fifteen runs of SAGE, from January 1990 to May 1992, yielded an average flux of 73 ± 17 SNUs, or 55% of the predicted 132 SNUs.

In these two gallium experiments and in Davis's chlorine

experiment one has to assume that the neutrinos one counts are primarily, if not exclusively, solar in origin. Experiments based on electron scattering of neutrinos, such as the Japanese Kamiokande II, don't have this limitation; they can determine the direction of the incident neutrinos. *More importantly, they detect all flavors of neutrinos, thus avoiding the ambiguity that Davis's experiment has* (see endnote 2.4).

An initial run of Kamiokande in 1987 yielded a flux of ^{8}B neutrinos—only 55% of that predicted by the standard solar model. With better calibration of the background, a second run ending in 1992 yielded 49%.

Thus all four detectors agreed that roughly half of the electron neutrinos predicted by the Standard Model are missing. However, the size of the deficit was different for each detector, presumably because each one samples a different part of the neutrino energy spectrum. But the most puzzling news came from the gallium detectors, which suggested that ^{7}Be production is severely depressed for some reason.

In order to sort out these quantitative disagreements, and to test the three-flavor explanation, three new neutrino detectors have been built: the Sudbury Neutrino Observatory, Borexino, and the Super-Kamiokande (see endnote 2.5).

In its first experimental runs the Super-Kamiokande confirmed previous estimates of the electron neutrino deficit, logging in at 47% of the prediction of the Standard Model. Then in 1998, scientists from the Super-K proudly announced they had demonstrated that electron neutrinos, produced in the earth's atmosphere by incident cosmic rays, have *mass*. You may remember that, according to the MSW theory, the oscillation of neutrinos among the three flavors can only occur if at least one has mass. Therefore, this experiment proved that the oscillation was a *possible* explanation of the deficit of solar neutrinos. However, the estimated neutrino mass, 0.05 electron volts, is too small to solve the dark matter problem of cosmology.

Following up on their discovery of neutrino mass, the Super-K scientists set up an experiment to test the MSW theory of solar neutrino oscillations. Since neutrinos change flavors more effectively in dense matter than in vacuum, they reasoned they could use the earth itself to modulate the flux of solar neutrinos. If the MSW theory works, some muon and tau neutrinos should switch back to electron neutrinos during the night, when they have to travel through the dense earth. Therefore, the flux of solar electron neutrinos should rise by night and fall by day.

They monitored the sun continuously for 504 days, beginning in April 1996. At the end of the run they had found *no* significant diurnal change in the electron neutrino flux.

Did this mean the MSW theory is dead? Not quite. The experimenters only claimed to constrain the circumstances in which the theory can work. What they showed by their experiment is that some ranges of the difference in masses between neutrinos are excluded.

Then, in June, 2001, experimenters at the Sudbury Neutrino Observatory made a stunning announcement: their measurements indicate that about half of all electron neutrinos leaving the sun arrive at the earth as tau or muon neutrinos. The MSW theory seems to be the answer to the long-standing deficit of solar neutrinos. Their conclusion is based on 1,169 events counted since April, 1999. More data is needed to confirm this result, and some details need to be clarified, but it looks as if we are home at last.

Incidentally, each of the three flavors of neutrinos has no more than 1/60,000 of the mass of an electron, far short of the mass needed to account for the missing mass in the universe.

At the present time theorists are engaged in reconciling the disparate measurements from the different neutrino observatories, with different energy thresholds and different capabilities. When and if that effort succeeds, we can fairly say that stellar astrophysics is set firmly on its foundations.

Chapter 3

The Deep Interior

nward and upward we float on our journey within the sun's interior. Eventually
we reach a point where the temperature
is too low (about 9 million kelvin) to sus-
tain the proton-proton chain. We have moved out of the
solar core, the region of energy generation, and into a new
region, the *radiative zone*.

Perhaps the first thing noticeable about this zone is that
the temperature, pressure, and density of the plasma are all
absolutely *steady*: this region is in a nice stable equilibrium.
Nothing boils, there are no dramatic flames, no violent
pulsations. Just a quiet steady climate of sheer hell! Physicists tell us this perfectly stable condition can only exist if
several separate kinds of equilibrium exist simultaneously.
In this chapter we'll survey the physical principles that prevail here.

Until recently this deep interior of the sun was solely the
province of theorists, who applied their most plausible
physical principles to construct mathematical models of the
region. In the past two decades the new technique of
helioseismology has been developed that allows experimentalists to observe this region indirectly and to compare their
results with theory. The comparison turns out to be very
satisfactory. In this chapter we'll survey the physical principles that rule here, and show how theorists applied these
principles to generate detailed mathematical models of
the structure of our star. In chapter 5 we'll examine
helioseismology.

The Balance of Opposing Forces

Every scuba diver knows that the deeper she dives, the
higher is the pressure. The reason for this effect, in both
the ocean and in the sun, is fairly obvious. The pressure the
diver feels at her depth arises from the weight of all the

water above her. The pressure at any depth must support the weight of water above that depth or else the ocean would collapse upon itself. Similarly in the sun the pressure at any level must support the weight of all the overlying plasma. This situation, a stable balance between pressure and gravity, is called *hydrostatic equilibrium*.

The plasma at these depths is about as dense as water. So it may seem surprising that the plasma can still be compressible and in fact behave more like a gas than a liquid. But the high temperature of the plasma here guarantees that the solar material cannot condense into a liquid or solid. This situation is fortunate since it allows theorists to apply the well-known laws that govern gases.

Gases are familiar objects to physicists. As early as 1690 Robert Boyle was experimenting with gases such as oxygen and nitrogen. When he compressed a fixed mass of gas, keeping its temperature constant, its pressure rose proportionally to the decrease of its volume. Later, in 1787, Jacques Charles confined a volume of gas and heated it. Its pressure rose in proportion to its temperature. These two relations can be combined into the so-called "perfect gas" law, which relates the pressure, density, and temperature of a fixed mass of gas (see endnote 3.1).

WHAT GOES IN MUST COME OUT: THERMAL EQUILIBRIUM

Outside the sun's core, the temperature is too low to generate thermonuclear energy. That means that all the energy the sun ever produces originates solely in the core. Then the energy must flow somehow toward the surface. What happens to it as it passes through all the intervening layers?

In principle the plasma around us could absorb some of it. In that case, the plasma would heat up. Or on the other hand the plasma here could be losing heat and cooling down. But we saw earlier that the plasma here is neither heating up nor cooling down; the temperature in every blob is stable. The blob is in "thermal equilibrium." This means that the stream of energy from the core flows right through this zone. Each blob of gas neither gains nor loses heat as the stream passes by.

Of course each little blob of plasma here contains a huge amount of heat, but it is insulated by all the layers above and below and doesn't need to be continuously heated to maintain its high temperature. It remains thermally stable while a feeble flow of energy passes through on its way to the surface (see endnote 3.2). (Actually the flow isn't all that feeble. Please recall that the sun loses *a million tons of mass a second* in the form of the sunlight that escapes at the surface!)

How can this energy "flow" from one layer in the sun to another? Three different mechanisms are available in principle: heat conduction, convection, and radiative transfer. When you boil an egg, the heat of the electric coil is first *conducted* through the bottom of the metal pot. In conduction, hot electrons in one part collide with cooler electrons in the adjacent part and pass along some of their kinetic energy as "heat." Next, the heat of the pot is *convected* to the egg by the circulation of bubbles of water. Bubbles of hot water rise, release their heat to their cooler surroundings, and descend for another load: a sort of internal conveyor belt.

When you remove the pot from the stove the red-hot electric coil warms your hands by the glow of radiation it emits. The heat is carried to your hands in little packets of light energy, the photons. *Radiative transfer* is fairly complicated in its atomic details; this chapter will focus on the easiest aspects, the action between photons and free electrons in the plasma.

All through the late 1800s most astrophysicists thought stars used convection to transfer heat, in analogy to the earth's atmosphere. Then, in 1906, Karl Schwarzschild, a professor at the University of Göttingen, tried out a different idea.

Schwarzschild was a child prodigy who published his first paper on the theory of orbits at the age of sixteen. As a young man, he was an avid balloonist, skier, and mountain climber. Later, he contributed to several fields of astrophysics, including stellar structure and spectroscopy. After Einstein formulated his general theory of relativity in 1916, Schwarzschild used the theory to show that if a star were compressed below a critical radius (now called the "Schwarzschild radius") the gravitational forces would become so large that nothing—not even light—could escape the star. For a star of the sun's mass, this critical radius is about 3 km. Schwarzschild had discovered black holes.

In 1906 however, Schwarzschild was studying the transport of energy through a star and using the sun as a test case. He postulated that radiation, not convection, determines the temperature in the outer layers of the sun. He also introduced the concept of radiative equilibrium, in which the total amount of energy flowing toward the surface is constant from one layer to another.

Schwarzschild was able to show that these assumptions led to the observed variation of brightness across the solar disk and also predicted the sharp edge of the solar disk. So he concluded that radiative transport must be more important in the sun than either convection or conduction. His result strongly influenced other astrophysicists, such as Arthur Eddington. But later on, Schwarzschild had to modify his sweeping conclusion. Convection, as we shall see, does play an important role in a particular layer of the sun (see endnote 3.3).

To summarize, the stream of energy from the core toward the surface is carried primarily by radiative processes, not by convection. Since photons do the heavy work of carrying the sun's energy, they deserve closer attention.

A CENSUS OF PHOTONS

The plasma here is so hot that the light it emits consists of soft X-rays, with wavelengths around a nanometer, or a ten-millionth of a centimeter. For comparison visible light has wavelengths around 500 nanometers. These X-rays consist of photons with energy ranging from about 400 to 1,300 electron volts. To understand the situation more clearly, we need to sort the photons according to their energy.

Suppose we select a small volume of plasma and count the number of photons with energy between say 400 and 410 volts, and then between 410 and 420 volts, and so on. When we finish, we plot the results and find the curve shown in Figure 3.1. A physicist would recognize this curve as that of a "black body."

A black body is one that absorbs all the light incident on it. The walls of a hot kiln are good practical examples. They radiate toward each other and absorb all the light they receive. If the kiln is kept at a uniform steady temperature, it becomes filled with radiation that has a unique spectrum, so-called "black body radiation." The spectrum doesn't depend on the wall material or the shape or size of the kiln, and only depends on the wall temperature—a remarkable

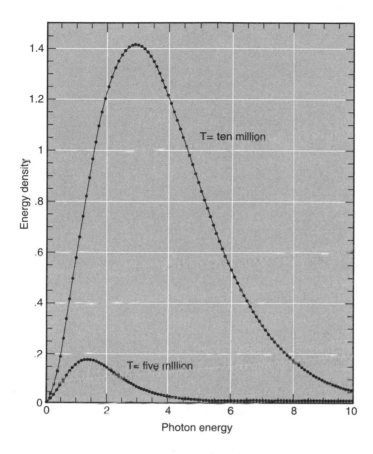

Fig 3.1 The energy distribution of photons in black body radiation. The density of radiative energy increases sixteenfold as the temperature of the plasma rises from 5 to 10 million kelvin. At the same time the peak of the curve shifts to shorter wavelengths.

property that implies a fundamental physical principle.

If we counted photons at places where the temperatures differ markedly (see Figure 3.1) we'd discover two remarkable characteristics of the black body spectrum. As we move from cooler to hotter regions, the peak of the spectrum shifts toward higher photon energies, or equivalently, to shorter wavelengths. In fact, according to Wilhelm Wien's law, the product of the temperature and the wavelength of the peak is a constant.

So, for example, at 10 million kelvin the peak lies at about 0.25 nanometers, while at 5 million it shifts to 0.50 nanometers. According to Wien, the temperature of the region is the *sole factor* that determines the peak's wavelength. So the hotter the plasma, the shorter the wavelengths of the light it emits. (This is true of many solid bodies, too. As we heat up a poker, its color changes from a dull red to yellow to blue-white.)

The two curves in Figure 3.1 are drawn to the same scale.

Notice how much larger is the area under the curve corresponding to 10 million kelvin than under the 5 million kelvin curve. In fact, according to the Stefan-Boltzmann law, the total emission of a black body increases as the *fourth power* of the temperature. So as we move from a region at 5 million kelvin to one at 10 million, the amount of radiative energy present in a cubic centimeter (the radiation "density") increases by a factor of *sixteen*.

In order for an enclosure like a kiln to contain black body radiation the enclosure must be held at a constant *uniform* temperature. Since the sun's temperature obviously varies from center to surface, the radiation at each place can only approximate a black body's. But the approximation is pretty good. The sun's temperature drops in 700,000 km from 15 million kelvin at the center to effectively zero at the surface, for an average gradient of a mere 21 kelvin per kilometer.

Nevertheless, you may ask, how can each little blob of plasma contain black body radiation corresponding to its local plasma temperature? After all, the existence of black body radiation would seem to require an enclosure of some kind and nothing here inside the sun remotely resembles an enclosure.

If we follow individual photons in the vicinity of our laboratory, we soon learn why. Photons can travel only a short distance (their "mean free path" is a fraction of a centimeter) before being absorbed: the plasma is very *opaque* to soft X-rays. This means that each blob can only "see" a very short distance into its surroundings and can exchange photons directly only with a few near neighbors. Since the average temperature gradient is 21 kelvin per kilometer, the temperature changes by less than a *thousandth of a kelvin* over a photon's mean free path. Each blob is effectively "enclosed" by neighbors that have nearly its own temperature and so its radiation resembles a black body's.

In the early days of stellar research, theorists had no clear ideas about the emission and absorption of radiation inside a star. Since black body radiation doesn't depend on any specific mechanism, but is a general consequence of the thermal equilibrium of an enclosure, it was a very attractive (and successful) assumption for theorists to make in building mathematical models.

In the nineteenth century several eminent physicists, including Boltzmann, William Thompson, James Jeans, and James Clerk Maxwell, struggled to predict the shape of the black body spectrum using classical physics. Although they found solutions for parts of the spectrum, they all failed to find the complete answer. Only after introducing the unfamiliar (and unpalatable) concept of discrete packets of light ("quanta") was Max Planck able to accomplish this task. He thereby reluctantly launched the quantum theory of light and matter.

THE FLOW OF RADIATION

So far we've learned that the radiation at any place has a black body spectrum that corresponds to the local temperature. But what determines the local temperature? Or more to the point how does the temperature vary from one place to another?

The energy from the core flows through this region as radiation. But the plasma is highly opaque to radiation. Somehow, the temperature gradient at any place must adjust so as to *drive* the radiation through the obstructive material. The situation is analogous to the flow of water through a clogged pipe (see Figure 3.2).

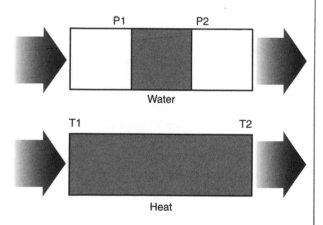

To push the water through the plug, the pressure must be higher upstream than downstream. In other words, a pressure *gradient* is needed. The larger the pressure gradient across the plug, the larger is the flow through the plug. By the same argument, the larger the plug's resistance to flow, the smaller is the rate of flow.

Now in the sun, the *temperature gradient*, not the pressure gradient, drives the flow of energy through the resistant (opaque) plasma. The rate of energy flow is fixed, once and for all time, by the nuclear generation processes in the core. That amount of energy *must* escape or the sun will explode. On the other hand, the opacity of each layer, which resists the flow, is determined by that layer's temperature and density. The sun has to adjust the temperature in each layer in

Fig. 3.2 An analogy between the passage of water in a pipe and the flow of radiation in the sun. A pressure difference (P1 is greater than P2) forces water through a clog in the pipe; a temperature difference (T1 is greater than T2) forces radiation through the opaque plasma.

such a way that the *gradient* across each layer is sufficient to drive the flow of energy through the opacity of that layer. It manages somehow to solve this complicated problem without the aid of a computer.

BORN-AGAIN PHOTONS

Although the total *amount* of radiant energy flowing upward toward the surface (the luminosity) does not change outside the core, the *quality* of the energy changes. As we float toward the surface the temperature drops and as a result the peak of the black body spectrum shifts toward longer wavelengths according to Wien's law. So the cooler the plasma, the fewer blue photons and the more red photons it contains. Moreover, since the energy E of a photon is related to its wavelength $E = hc/\lambda$ where h is Planck's constant and c is the speed of light in vacuum) more red photons are needed to carry the same amount of energy. How are new photons created?

Suppose that you could follow the wanderings of one energetic photon and its descendents. In Figure 3.3, the photon's

Fig. 3.3 Photons diffuse through the sun in a random walk, occasionally splitting into two daughter photons.

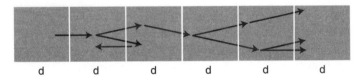

d　　　d　　　d　　　d　　　d　　　d

path has been partitioned into intervals of a mean free path. The photon heads upward (at the speed of light within the plasma), out of the cell where it was born and into the next higher one. There, by definition of a mean free path, it is "absorbed" and disappears as heat into the plasma. After a short interval, the plasma may emit another photon identical to the original one and this one might return to the first cell or head outward into the third cell. In such a chain of events the photon is in a sense "reborn" after each absorption and wanders randomly, one mean free path at a time in either direction. Such a path is called a "random walk."

This name arose from the anecdote about the drunk who stands under a lamppost on a sidewalk, wondering which way home lies. He takes single steps to the right or left along the sidewalk, in random order. An analysis of his progress shows that his average position always remains at the lamppost but that his mean distance from the post (in either direction) increases as the square root of the number of steps he has taken.

But let's return to the single photon. If nothing else interfered, the reborn photon would wander further and further in either direction from its original cell. But at least two things do interfere. First, the mean free path is shorter the deeper it wanders (because the mass density is higher there) and longer the higher it wanders. This will bias its net progress upward.

But more important, it is far more likely that this photon will not be reincarnated as described above. It is more likely that once it enters a cell that is cooler than the one it started from, two daughter photons will be emitted in its place. In the cooler cell the black body spectrum contains a tiny percentage fewer of energetic photons so the probability of a reincarnation is (very slightly) reduced. The probability of emitting less energetic daughter photons is increased. The sum of the daughter's energies will equal hers but they will each be less energetic ("redder"). They too will wander randomly and they too will have daughter photons eventually.

This scenario is very crude, but it emphasizes the gradual degradation of high-energy photons, emitted in hot depths of the radiative zone, to more but lower energy photons emitted higher in the zone. The net transfer of energy upward is partly the result of the random wandering of photons in the plasma, with step sizes decreasing inward and increasing outward. This is a *diffusion process* similar to the spread of a drop of ink in water. If you were to follow all the generations of daughter photons you would find that it takes about a million years for the last generation to reach the surface, where it escapes into space as sunlight.

In this description of photon wandering, the physical processes that create and destroy photons were glossed over. It's time to examine these more closely.

The Birth and Death of Photons

Photons are created and destroyed as they interact with the charged particles in the plasma. Consider the simplest interaction: the emission and absorption of photons by free electrons.

Protons and electrons in the plasma swirl and dodge in a complicated ballet. Their opposite electric charges tend to pull them together, but the electrons move too fast to be captured. They swerve past the protons in tight trajectories, constantly changing speed and direction. In other words they are constantly *accelerating*.

Now, according to Maxwell's electromagnetic laws, an accelerated electric charge, like an electron, emits light. You can understand this statement qualitatively if you look at Figure 3.4.

Fig. 3.4 An accelerated electron emits a wave of electromagnetic energy, a photon of light.

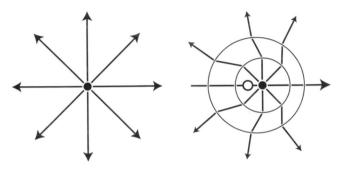

Imagine that you are following a single electron as it moves in a straight line at constant speed. Its electric field stretches out radially in all directions. If now the electron *accelerates* in the same direction, its surrounding field must change accordingly. Maxwell showed that this change cannot occur instantaneously at all distances, but must propagate as a wave at the speed of light. The right side of Figure 3.4 is a snapshot of this electromagnetic wave moving outward, to signal the far distant field that the electron has changed speed. Notice the sheared lines of force in the wave. The wave represents a pulse of energy that the electron has lost. The electron tends to slow down as a result. In classical terms the wave is a light wave; in quantum terms the wave is a photon.

So as electrons accelerate randomly in the plasma, they emit light. The *inverse* process (absorption) also occurs. If we reverse time in Figure 3.4 a wave will *converge* on an electron and give it a kick. Its speed increases and it goes its merry way. From the microscopic point of view, its increased kinetic energy represents additional *heat* that the photon has added to the plasma. Within a very short time, this electron will collide with another and share its enhanced kinetic energy with its partner. In a chain of collisions the excess energy is redistributed among all the electrons. In this way the photon energy absorbed in individual encounters is rapidly distributed throughout the plasma. This process of absorption is called a "free-free" absorption, because the electron moves freely before and after the interaction.

At first glance, the charged particles of the plasma appear to be moving chaotically. But there is an overall order to their motions, which we now consider.

A POPULATION CENSUS OF PARTICLES

Let's zoom in now on the plasma inside a small cube. Although the plasma in the cube is not flowing as a whole, the electrons, protons, and heavy ions within it are moving rapidly and randomly, colliding frequently with each other. So frequently in fact that their mean free paths are much shorter than the width of the cube.

Now, take a census of all the electrons in the cube, sorting them according to their kinetic energy. Count the electrons with kinetic energy between, say, 1.0 and 1.1 kilovolts, and then between 1.1 and 1.2, and so on.

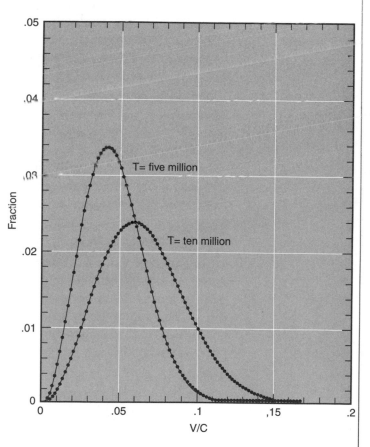

When you finish, plot the results. Figure 3.5 shows the result you would get: the fraction of all the electrons in your sample that have a specified energy.

This distribution of particle energies was predicted by James Clark Maxwell in 1860 and independently by Ludwig Boltzmann, cofounders of the science of statistical mechanics. *Any collection of particles, colliding within an enclosure that*

Fig. 3.5 The Maxwell-Boltzmann distribution of electron energy in a plasma. As the temperature of the sample rises from 5 to 10 million kelvin, the curve shifts to higher energies.

is maintained at a uniform temperature T, will tend toward a Maxwell-Boltzmann distribution of energies.

The kinetic energy spectrum is peaked, like the black body spectrum. Like the black body spectrum the peak lies at an electron energy directly related to the temperature, $3/2\ kT$, where k is the Boltzmann constant. Therefore, as we move from a region at 5 million kelvin to 10 million, the peak shifts toward higher energy. However, since the area under each curve represents the *fixed* number of electrons in your sample, the areas are equal.

You can understand why most electron energies cluster near the average. If two randomly moving identical particles collide, the most probable result is that the faster one shares its energy with the slower one, and both then move toward an average energy.

This law of energies applies equally well to all the other particles that make up the plasma. If now you sort all the protons, for example, you'd find exactly the same curve as for the electrons. The average proton energy would also be $3/2\ kT$. There is a difference, of course: since a proton is 1,836 times heavier than an electron with the same kinetic energy $(1/2\ mv^2)$, it moves more slowly.

The beauty of the Maxwell-Boltzmann distribution is that it doesn't depend upon the particular kinds of forces that act in particle collisions, or upon the mixture of different nuclei. As long as collisions are frequent enough (i.e., if the particle density is sufficiently high) they will maintain the distribution indefinitely. Although collisions constantly change each individual particle's energy, the distribution of energies among the large number of particles is maintained.

Fig. 3.6 Two different distributions of energy among 12 different electrons. The total available energy is supposed to be the same in each case.

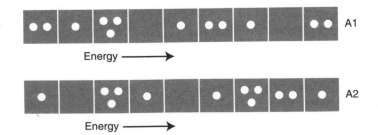

The Maxwell-Boltzmann distribution is so robust because it describes *the most probable situation* among a collection of interacting particles (for details, see endnote 3.4).

Boltzmann was a brilliant but tragic figure in the history of physics. He pioneered in introducing statistical arguments about the motions of gas particles and thus was able to explain the macroscopic laws of thermodynamics in terms of the microscopic motions of atomic particles. But his work in this new science of statistical mechanics was severely criticized by his peers who abhorred the idea that laws of probability underpin the classical laws of heat. Ill and depressed, Boltzmann took his own life in 1906, shortly before the experiments of Jean Perrin on the random motions of dust particles (Brownian motion) validated his work.

Maxwell was only forty-eight when he died, but in his short life he revolutionized physics. He is best known for his four equations of electromagnetism, for his prediction that light is a form of electromagnetism, and for the Maxwell-Boltzmann law. But early in his career he flirted with astronomy: he showed that the rings of Saturn had to be composed of small solid particles.

THERMODYNAMIC EQUILIBRIUM

So far we have found that each volume in the radiative zone contains photons with a black body spectrum and particles with a Maxwellian energy spectrum. Both spectra are characterized by the same local temperature, T, which determines their average or most probable energies. This condition (a state of "thermodynamic equilibrium") cannot be accidental. There must be good reasons why it prevails. After all, the radiation and the particles interact, exchanging energy by absorption and emission processes. (We illustrated one of these, the free-free process.) Somehow the black body radiation constrains the particles to move according to the Maxwellian distribution and the Maxwellian particles generate black body radiation—a perfect symmetry in which the two energy distributions are *consistent and reinforcing*.

How does this happen? This is the theoretical problem that Max Planck solved in 1896. To solve it he had to make a radical break with the past.

He imagined a box with black walls that was kept at a constant temperature. Inside the box he imagined a set of "oscillators"—electrons that vibrate back and forth to produce radiation of different wavelengths. Since the walls are black and opaque, no radiation can escape from the box.

He realized that if an oscillator loses kinetic energy by radiating, the radiation in the box must gain some, and vice versa. This simple idea gave Planck the principle he needed to work out how much radiant energy the box contains, at least in terms of the energy of a single oscillator. His big problem was to find the energy of a single oscillator.

In order to keep the math simple, Plank proposed that an oscillator could only accept energy in finite amounts, which later were called "quanta." This critical assumption allowed him to predict the observed frequency distribution of black body radiation, the so-called Planck law. His success launched one of the greatest ideas in the history of physics: quantum mechanics.

But oddly enough, Planck was never convinced that energy comes in finite packets, or quanta. He thought his derivation of the correct formula was somehow incomplete and "alien to the spirit of electromagnetic theory." He spent much of his later life trying to eliminate the alien photon from physics. But obviously photons were here to stay. Some years later, Albert Einstein and Satyendra Bose derived Planck's law far more rigorously, by applying statistical mechanics to photons.

Fig. 3.7 The emission spectrum of an electron oscillator. Instead of radiating only at its resonant frequency (v_0) it emits over a wider range of frequencies.

In endnote 3.5 we answer the question we raised earlier: why does the radiation in an isothermal enclosure have a black body spectrum and why do the particle energies have a Maxwellian distribution? The answer (or one answer) is

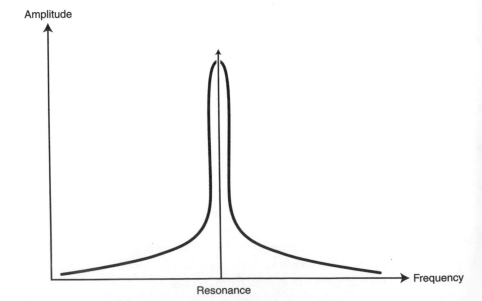

that each of these energy distributions is the *most probable* one. As long as the particles can exchange energy among themselves (through collisions) and as long as the particles can exchange energy with the radiation field, these two energy distributions will be maintained and will reinforce each other.

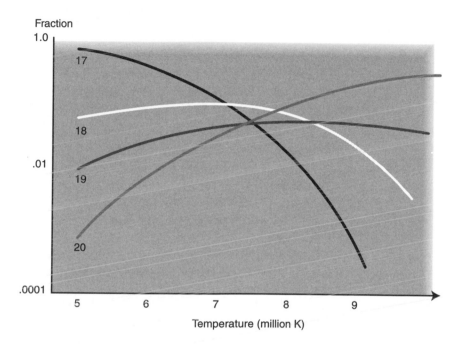

So far we have ignored all ions heavier than helium. In endnote 3.6 we take a look at the rest of the constituents of the plasma.

MODELS, OLD AND NEW

What does all this mean in terms of constructing a mathematical model of the sun? We have surveyed the physical principles that govern the equilibrium of plasma in the solar interior. We found first of all that *hydrostatic* (or mechanical) equilibrium of the sun requires a definite relationship between the pressure, plasma density, and gravity at each position. Second, *thermal* equilibrium requires a particular temperature distribution in order to maintain a constant flow of radiant energy from one layer to another, despite the high opacity of the plasma. Finally, *thermodynamic* equilibrium arises from the exchange of energy between particles and photons, and in turn insures that certain very general laws (black body radiation, the Maxwellian

Fig. 3.8 As the temperature rises, iron atoms lose more and more of their bound electrons. Between 5 and 10 million kelvin, ions that have lost 17, 18, 19, or 20 electrons exist in different proportions.

velocity distribution, and the Boltzmann distribution) are followed.

In the core, the proton-proton chain generates the solar luminosity. This thermonuclear energy is carried from the core to the surface in the form of a net flux of black body radiation. Finally, despite the high density of the plasma here, the perfect gas law relates the pressure, temperature, and density.

With these principles in hand we could now sit down and construct a mathematical model of the sun, which would predict the temperature, density, and pressure at each interior point. We are in fact in a better position to do this than theorists of the late nineteenth and early twentieth century. It is worth taking a moment, however, to recall the progress these pioneers made despite their relative ignorance of several crucial elements.

The first to construct such a model was J. Homer Lane, an obscure American engineer based in Washington, D.C. In 1869 Lane applied the principle of hydrostatic equilibrium (which was well known to meteorologists) to a gaseous sphere, and assumed that heat was transported internally by convection. In such a situation the pressure p and density ρ are related by a simple law $p = C\rho^{\gamma}$, where the exponent, γ, equals 5/3 for a monoatomic gas. In addition, Lane made the bold assumption that the solar interior behaves as a perfect gas, despite its mean density 1.4 times that of water.

With these three constraints in hand, Lane predicted the variation of the sun's temperature from the core to the surface, and tried to test his model by comparing his prediction of the surface temperature with observations. Unfortunately the latter were so imprecise as to draw any definite conclusion. However, Lane obtained one result that was crucial for later work on the evolution of stars, namely, that if a perfect gas star contracts gravitationally, its internal temperature *rises*.

Lane's methods were extended by several other scientists, including William Thomson, later Lord Kelvin, but the next important step was taken by Robert Emden, an assistant professor of physics and meteorology at the Technische Hochschule in Munich.

In 1907 Emden published a textbook for his students which

incorporated astrophysical examples that built on the work of Lane. He investigated a series of stellar models in which the pressure and density were related by a simple mathematical relationship, such stars being known as "polytropes." Perhaps his most important contribution was a discussion of the conditions necessary for a star to have a finite radius. His models therefore began to resemble real stars. The extensive numerical tables in his textbook proved invaluable to other theorists, among them Eddington.

Sir James Jeans took a dim view of such model-building. He was one of a small group of distinguished British scientists in the early 1900s and he argued that too little was known about the physics of stellar interiors to begin to construct models. In particular, since the laws governing the release of subatomic energy were completely unknown, stellar models could only be speculative. Jeans was arguing for the comprehensive approach to constructing models that has become feasible only after the proton-proton chain was discovered in 1939.

However, Arthur Eddington took the opposite point of view. In his monumental monograph of 1926, "The Internal Constitution of the Stars," he wrote:

> Now it is quite true that a theory of the rate of liberation of sub-atomic energy is a conceivable approach to the problem of stellar radiation. In the present state of our knowledge such theories are little more than guesswork and results are rudimentary. But it is unsound to argue that no other procedure is permissible. The amount of water supplied to a town is the amount pumped at the waterworks; but it does not follow that a calculation based on the head of water and the size of the mains is fallacious because it evades the problems of the pumping station.

In short, Eddington was willing to make some assumptions about the distribution of energy release and proceed from there.

Eddington's basic goal was to predict the luminosity of stars of different masses, using the principles Lane and Emden had pioneered. Giant stars with very low mean density had been discovered since Lane's work and this convinced Eddington that the perfect gas law could very well apply in their interiors. (He didn't think it applied to stars of lower mass, however.)

Moreover, he was persuaded by the work of Karl Schwarzschild that energy is transported primarily by radiation in hot stars, not convection. Recall that Schwarzschild had assumed that radiative equilibrium prevails in the solar atmosphere and had successfully predicted the variation of brightness across the solar disk.

Eddington was handicapped in several ways. First, he had no idea of the true chemical composition of stellar material. He assumed a composition similar to that of earth, which led him to adopt a mean atomic weight μ equal to 2. Since the luminosity of a star varies as a high power of μ, Eddington's results were sure to be compromised. Ironically, only a year after he published his massive work, Cecilia Payne established the predominance of hydrogen in stars, which implied that $\mu = 0.5$.

Then, like everyone else at the time, Eddington had limited information on the sources of opacity and essentially no idea of the sources of subatomic energy generation. Here is where his superb physical intuition and mathematical skill came to his aid. He recognized that the rate of energy generation was significant only in a small central core and varied according to some unknown slowly varying function of depth. Likewise he recognized that the opacity probably decreased in the core and varied slowly outside the core. He therefore made the bold assumption that the *product* of the two unknown functions *remained constant* throughout the interior of a star. This assumption greatly simplified his calculations.

Eddington was concerned with stars more massive than the sun. In these, the pressure of radiation, "like a wind blowing through the stellar interior" helps to support the weight of overlying layers. He was the first to incorporate radiation pressure in a stellar model.

Using the principles of hydrostatic equilibrium, radiative equilibrium, and thermodynamic equilibrium, Eddington was able to construct models for stars ranging in mass from a half to forty solar masses, and to predict their luminosities, except for a single unknown constant related to the unknown energy generation and opacity functions.

His results revealed several important conclusions. First and foremost, *the luminosity of a star is proportional to the cube of its mass*. In Figure 3.9 we see a comparison of his prediction and the data available to Eddington. The fit is quite re-

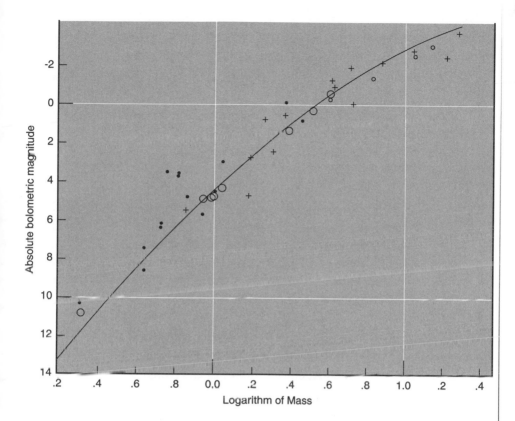

markable considering the state of his ignorance at the time. Clearly some of his uncertainties tended to cancel out in his calculations. Recent calculations, employing more complete physics of the absorption coefficient K, indicate that luminosity varies approximately as the fourth power of the mass.

Fig. 3.9 Eddington's mass-luminosity relation compared to observations of his time. (Courtesy of Dover Publications)

Secondly, in contrast to the luminosity, the radius of a star depends only weakly on its mass (approximately as the sixth root of the mass!). Third, the central temperature of a star varies inversely as the square root of its mass. Here Eddington's uncertainties show up most vividly: he predicted a central temperature of 40 million kelvin for the sun, in contrast to the modern value of 15 million.

In summary, Eddington's investigation, despite its flaws, demonstrated that the three principles he employed must be very close to reality. In particular, radiative equilibrium prevails through most of the interior, especially in stars hotter than the sun. His insights were crucial to the further development of the subject.

Eddington has been described as "the most distinguished astrophysicist of his time." As such he was a fierce competitor, especially against Jeans, with a keen sense of his own worth. He also had his human side, however. He enjoyed biking and kept detailed records of his mileage. He was an avid crossword puzzler, who could knock off a Times puzzle in a few minutes. And as a devout Quaker he absolutely refused to kill, and became a conscientious objector during the first World War.

In his later years, Eddington became one of the principal proponents of Einstein's general relativity theory. He eventually tried to unite quantum mechanics and relativity in a "fundamental" theory. However, in constructing the theory he was guided by some ingenious (but unsubstantiated) manipulations of the constants of nature, such as the speed of light and Planck's constant. For example, he claimed to deduce the number of particles in the universe from the ratios of certain atomic constants.

> I believe there are 15,747,724,136,275,002,577,605, 653,961,181,555,468,044,717,914,527,116,709,366, 231,425,076,185,631,031,296 protons in the universe and an equal number of electrons.

His theory was met with a storm of criticism. Eddington became increasingly dogmatic and defensive. "One should never believe any experiment until it has been confirmed by theory."

In 1920, he gave his views on the place of speculation in science, which still make good reading:

> If we are not content with the dull accumulation of experimental facts, if we make any deductions or generalizations, if we seek for any theory to guide us, some degree of speculation cannot be avoided. Some will prefer to take the interpretation which seems to be most immediately indicated and at once adopt that as an hypothesis; others will rather seek to explore and classify the widest possibilities which are not definitely inconsistent with the facts. Either choice has its dangers: the first may be too narrow a view and lead progress into a cul-de-sac; the second may be so broad that it is useless as a guide and diverges indefinitely from experimental knowledge. When this last case happens, it must be concluded that the knowledge is not yet ripe for theoretical treatment and speculation is premature. The time when speculative theory and observational research may profitably go hand in hand

is when the possibilities—or at any rate the probabilities—can be narrowed down by experiment, and the theory can indicate the tests by which the remaining wrong paths may be blocked up one by one.

How to Build a Star

Eddington's investigations of stellar structure led to a remarkable general principle, stated most clearly by Henry Norris Russell, the famous astrophysicist at Princeton University:

> The luminosity of a star in radiative equilibrium is determined solely by the star's mass and chemical composition.

Take a moment to consider what this means. A star forms out of an interstellar cloud with a certain mass and chemical composition. If it obeys the three principles described above (hydrostatic, radiative, and thermodynamic equilibrium) and also obeys the physics of thermonuclear generation, then its final internal structure is completely determined. In particular its final luminosity is determined and is a function (something like the fourth power) of its mass.

Over the past fifty years, astronomers have developed increasingly sophisticated numerical techniques to compute the radial distributions of temperature, density, and composition within a stable star, and indeed to chart their evolution over billions of years. These calculations employ all the physical principles we have discussed. In accord with Russell's principle, once a definite stellar mass and initial composition are chosen, a star's complete internal structure can be determined. A final temperature-density model for the sun is considered acceptable if it reproduces the observed solar luminosity and radius within the known age of the sun.

In Figure 3.10 we see the results of a recent model of the solar interior, constructed by John Bahcall and Roger Ulrich to investigate the production of solar neutrinos. It employs the best physics available today. As described in chapter 2, such a model alone is inadequate to predict the observed neutrino flux, and must be augmented with new particle physics. However, chapter 5 will explain how astronomers use helioseismology to check other details of this model, and find, to their delight, excellent agreement.

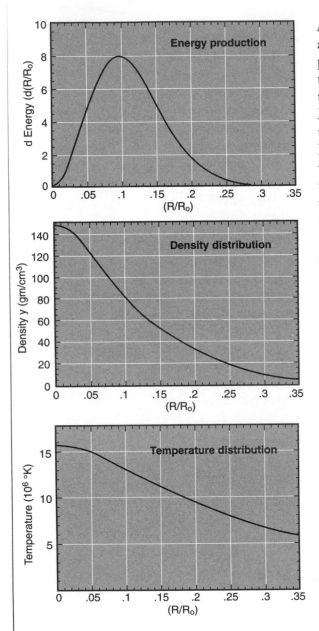

All through this chapter we've avoided any mention of plasma flows in the solar interior. In the radiative zone, this is not a serious omission. Most evidence suggests that this zone is very nearly static. In the next chapter, however, we'll discover a large region—the solar convection zone—in which flows are essential for transporting heat to the surface.

Fig. 3.10 The Bahcall-Ulrich model of the solar interior, based on the fundamental principles described in the text. The three panels show energy production, plasma density, and temperature, all as functions of solar radius. (Courtesy of Reviews of Modern Physics)

Chapter 4

ORDER AND CHAOS: THE CONVECTION ZONE

e have come a long way in our journey from the sun's center—almost 500,000 km. At this point, the temperature has fallen to about 2 million kelvin, and the plasma is about as dense as styrofoam. Nothing moves, nothing is remarkable. There is nothing much to comment on. Soon we expect to hit the solar surface.

But life has its surprises! A little higher up, we pass through a sharp boundary and here the plasma is definitely tumbling, slowly but unmistakably. We have left the sun's radiative zone and entered its convection zone.

The first order of business is to survey this new region. Cruising rapidly upward, we learn that the zone is over 200,000 km thick and extends nearly to the solar surface. Within it, the plasma is moving in complicated patterns. There are huge streams of hot plasma, 100,000 km across, rising majestically at a few centimeters per second. Then at other places between these hot updrafts we find narrow plumes of cooler plasma that plummet toward the bottom of the zone at a few meters per second (see Figure 4.1). Evidently a slow circulation of

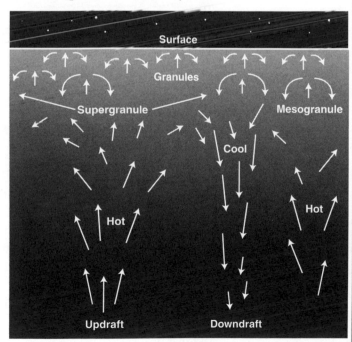

Fig. 4.1 Flows inside the convection zone. Some of the special convective cells, detectable at the surface, are labeled.

Fig. 4.2 Supergranules covering the solar surface. Plasma flows horizontally inside a cell and sinks at its edges. The cells are typically 30,000 km across and last for a day. (Courtesy of the National Solar Observatory)

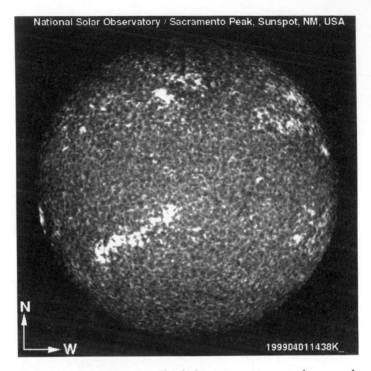

National Solar Observatory / Sacramento Peak, Sunspot, NM, USA

199904011438K_

plasma is occurring in which heat is transported upward. Hot streams rise, discharge their loads of heat to the cooler surroundings, spread out, and then join the downdrafts to descend. This is the way convection works, one of the three principal means of transporting energy within a star.

A few thousand kilometers below the surface, the broad hot updrafts break up into a gaggle of convective cells. Plasma swells up at the centers of these bubbles and sinks at their edges. The largest cells are the so-called "supergranules" (Figure 4.2), about 30,000 km wide and only a few thousand kilometers thick. They flip over in about a day, at a few tenths of a kilometer per second, before disintegrating into smaller cells. (This may sound leisurely but remember that a tenth of a kilometer per second amounts to 220 miles per hour!)

Somewhat closer to the surface we find "mesogranules"— about 5,000 km in size—that take several hours to turn over. And just under the surface lie still smaller cells, the "granules." They are packed closely to cover the whole surface, in a tile-like pattern (Figure 4.3). Granules are about 1,500 km in size (about the size of Alaska but thousands of kelvin hotter!) and exist for only about eight or ten minutes before dissolving. These cells are so close to the surface that they lose their heat by radiating into space.

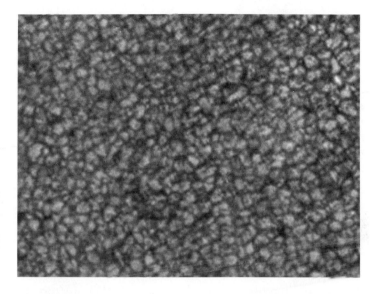

Fig. 4.3 Small convective cells ("granules") at the solar surface. The cells are typically 1,500 km in size and last for 5 minutes. (Courtesy of Big Bear Solar Observatory)

At higher magnification, the plasma's motion is much more complicated. It rolls and twists in an extremely chaotic manner, never seeming to repeat itself. Apparently the motions we see depend on the size of the volumes we examine. The big updrafts and the convective cells move in a fairly systematic fashion, as we just noticed. But imbedded in these large features are swirling eddies of all sizes down to the limit of our instruments. These small currents roil chaotically in *turbulent* flow.

We are faced with a blizzard of questions. Why does the sun need a convection zone at all? How can we account for all this complexity? Why do three sizes of cells predominate? How can convection be orderly on large scales and chaotic at small scales?

We start with the simplest question of all: why should a heated fluid begin convective motions? The answer lies in the laboratory.

How Convection Starts

In 1901 Henri Bénard, a French physicist, was doing simple experiments with warm liquids. He would fill a shallow pan with a very thin layer of oil or paraffin and slowly raise the flame under the pan. At first the liquid just conducted the heat to its top, without any visible motions. But eventually a critical temperature would be reached at the base of the layer, at which the liquid would begin to churn. The critical temperature depended strongly on the thickness of the layer and only weakly on the type of liquid. Very viscous fluids, like thick oil, settled into a characteristic and beautiful pattern of polygonal convection cells (Figure 4.4) that covered the whole surface. The flow in these cells is smooth, rising in the centers and falling at the edges.

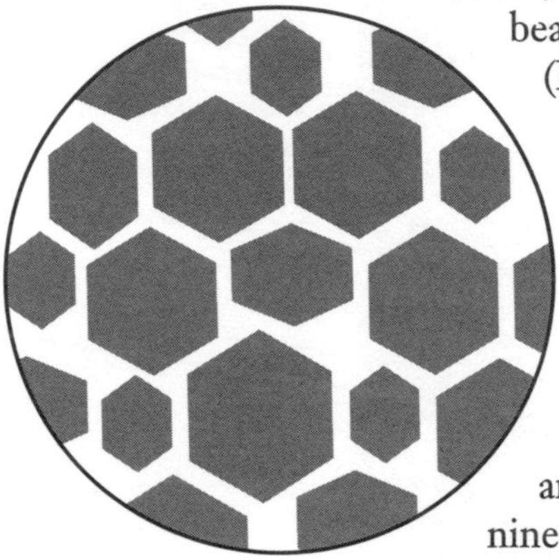

William Strutt, who would later be knighted as Lord Rayleigh, was fascinated by Bénard's little cells. He was one of the most creative British scientists of his day, and received the Nobel Prize in physics in 1904 for his many and varied contributions. Although only twenty-nine he answered a question that almost everyone has asked: "Why is the sky blue?" With a lot of mathematics he showed the lovely blue is caused by sunlight scattering off air molecules in the atmosphere. Later on, he and Sir James Jeans tackled the problem of black body radiation. Although known primarily as a theorist, he was also codiscoverer of the element argon, an inert gas that has practically no chemical reactions and was therefore difficult to detect. Almost as an exercise, this clever man devised a simple theory to explain Bénard's results.

Fig. 4.4 Polygonal Bénard cells form in a thin layer of viscous fluid, heated from below. Granules resemble them superficially.

He pictured a static layer of liquid heated from below (Figure 4.5). An isolated blob of liquid near the bottom of the layer heats up and expands. Like a hot air balloon, it becomes buoyant and tends to rise. But the liquid around it is viscous and tends to hold it back. If the buoyant force exceeds the drag, the blob will rise. Whether it will continue to rise depends on how quickly it cools by conducting heat to its surroundings. If the blob loses its heat too quickly, it loses buoyancy and hardly rises at all.

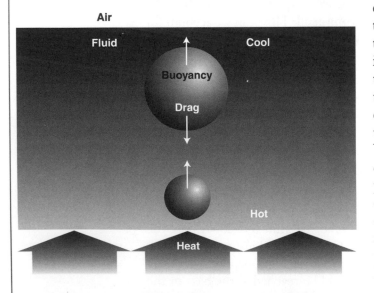

Fig. 4.5 Lord Rayleigh's picture of the situation inside a Bénard layer. A hot bubble will rise if the buoyant force equals or exceeds the drag force.

So the onset of convection depends on the competition among buoyancy, viscosity, and heat loss by conduction. Each of these factors is characterized by a numerical coefficient. Each kind of liquid (or gas) has its own coefficients of viscosity, heat conduction, and expansion. Rayleigh combined them in a dimensionless ratio, now called the *Rayleigh number* in his honor, which also involves the temperature gradient across the layer as well as its depth and the strength of gravity. Any combination of coefficients and temperature gradient that produces a *critical* value of Rayleigh's number implies that convection will start in the layer.

In laboratory experiments like Bénard's, the critical value of the dimensionless Rayleigh number is 1100. In comparison, the Rayleigh number in the solar convection zone is about 10^9! All this huge number tells us is that convection in the zone is extremely vigorous, far beyond the conditions for convection to begin. But the physical principles that govern convection are still valid.

WHERE CONVECTION STARTS IN THE SUN

Karl Schwarzschild, the astronomy professor at Göttingen whom we met in the last chapter, derived an alternate criterion for the onset of convection, which is better adapted to stars than is Rayleigh's. After all, stars consist of thick layers of hot radiating gases, not shallow layers of warm liquids.

As we learned earlier, the temperature gradient in each layer of the sun has to adjust so as to drive the flood of radiant energy toward the surface through the opaque plasma. As we approach the surface the plasma becomes increasingly opaque, because, as the temperature falls, heavy ions can retain more and more of their electrons. These ions then block the flow of photons more effectively.

So as we approach the surface the sun's negative temperature gradient becomes steeper and steeper. But there is a limit to how steep the gradient can get. Beyond that limit, radiative transport of energy is no longer feasible and convection takes over the job of shuttling heat toward the surface.

Schwarzschild worked out the critical temperature gradient at which convection takes over. He wondered what would happen to an isolated bubble of gas in the radiative zone if it happened to rise a bit. As it rises into lower pressure gas, it must expand and therefore cool. Outside the bubble, radiative transport determines the temperature gradient. But *inside* the bubble, the temperature is determined by two factors: its cooling due to expansion, and any additional cooling due to a loss of heat to its surroundings.

Suppose, Schwarzschild said, that *no* heat is lost, that the bubble expands "adiabatically." If the temperature gradient outside the bubble is steep enough, the bubble will always be hotter than its surroundings and it will continue to rise buoyantly, like a hot air balloon. As it moves upward, the bubble's temperature will change according to a so-called "adiabatic gradient."

Schwarzschild concluded that convection appears in a star wherever the radiative temperature gradient is steeper than the corresponding adiabatic gradient. In that situation, convection starts spontaneously, as a more efficient means of transporting energy.

THE BUBBLE THEORY

Schwarzschild's criterion only says what part of a star can no longer transport the flux of energy solely by radiation, and must convect energy. The criterion is not a *description* of how convection works. During the past sixty years astrophysicists have worked hard to develop a realistic theory of stellar convection, based on sound physical principles. It is worth recalling the first breakthrough in this effort, and then some recent advances.

As recently as the 1930s, astronomers who were building mathematical models of stars had no realistic means of treating convective motions. In particular they had no way of estimating how well and how far a hot bubble would carry its load of heat. So they avoided describing the details of convective motions by assuming, reasonably enough, that convection simply drives the stellar temperature gradient to the critical adiabatic temperature gradient throughout the convection zone. That assumption sufficed for their purpose.

Then the Germans picked up the problem. A series of studies in the mid-1930s, first by Ludwig Biermann, then by Heinrich Siedentopf, and Albrecht Unsöld, produced a crude description of heat transfer by rising and falling bubbles. Their work was extended and applied to more realistic stellar models in 1953, by Erika Böhm-Vitense, one of the few female astrophysicists of her time.

The basic idea in Biermann's approach was the "mixing length," which is defined as the average distance a blob rises before merging with its surroundings. Actually, Biermann borrowed this concept from his colleague in Göttingen, Ludwig Prandtl, who was a world-famous hydrodynamicist. Prandtl saw an analogy between the mean free path of a molecule and the distance a convective cell travels before disintegrating. Biermann used the mixing length idea to construct a crude model of rising and falling bubbles.

His simple picture was extended by Böhm-Vitense, who calculated the details of heat transport: how fast the bubbles would rise and how much heat they could transport. It was a distinct advance. The only missing factor in this whole scheme was the size of the mixing length itself, which nobody could predict. She guessed it must be related to the local pressure scale height, the distance in which the plasma pressure decreases by a factor of e (approximately 2.72). But in practice the mixing length was left as a free parameter, which could only be determined empirically by comparison with observations. A small multiple of the pressure scale height seemed to work reasonably well, but nobody knew why.

CONVECTION IN A BOX

During the following three decades (1955–85) astrophysicists were able to advance beyond the mixing length theory of convection. But along the way, they were forced to adopt a variety of simplifying assumptions about the behavior of plasmas. One reason why they couldn't solve the problem once and for all time, is that the equations that govern the motions of convecting plasmas are *nonlinear*. That means, for example, that a change in the velocity at some position in the plasma depends on the product of the local velocity and the gradient of velocities in the neighborhood. Such equations cannot easily be solved by the classical methods of mathematical analysis.

In the past decade, however, the availability of cheap, fast computers with lots of memory changed the whole approach to theory. It was no longer necessary to make severe assumptions—one could put in all the basic physics and allow the computer to calculate the evolution of the flows. This method is called numerical simulation and the present generation of astrophysicists has become quite accustomed to simulating nature.

The basic approach is to set up a conceptual three-dimensional grid of reference points within a rectangular box that is filled with plasma. The temperature at the top and bottom of the box is fixed and the sun's gravity is assumed to act on the plasma. Because the temperature decreases toward the top, convective motions begin immediately. Heat begins to flow through the bottom of the box and is radiated away at the top. The motion of the plasma at every grid point can be calculated, for each interval of time, with the aid of the three basic equations of hydrodynamics. These equations express the following principles:

1. Plasma is neither created nor destroyed in the box;

2. Energy (thermal plus kinetic plus gravitational) is neither created nor destroyed in the box;

3. The change of momentum of each blob in the plasma (the product of its mass and vector velocity) is determined by the various forces that act on it.

However, the properties of the real solar convection zone change radically from top to bottom. The temperature drops from 2 million to 7,000 kelvin and the pressure drops by a factor of a hundred. Even the most powerful computers available today can't follow the detailed motions of the plasma over such a huge range of variables. So theorists can only investigate convection in isolated parts of the convection zone rather than over its full extent. Some simulations investigate the deep parts of the zone where the typical sizes of cells are tens of thousands of kilometers. Other simulations examine convection near the top of the zone where convective cells are only a few thousand kilometers in diameter.

Unlike the computational box, the real sun has no walls. To get around this problem, theorists impose "periodic" boundary conditions. In effect they imagine the solar convection zone is made up of a large number of identical boxes, set side by side.

The theorist also has to suppress sound waves in his box. Rapid changes of pressure that occur in a convecting fluid naturally generate sound waves (just as a boiling pot of water produces plopping sounds). To follow each little sound wave would require far more computing power than the theorist has. So he artificially boosts the viscosity of the plasma in his box to damp out the noise.

But the worst limitation of simulations is the difficulty of including all the extreme conditions inside the sun within one realistic calculation. Nevertheless, real insights have been attained with even limited calculations. Let's review some of these results. We begin with the situation close to the solar surface. In this region, Bob Stein, an astrophysicist at Michigan State, is an expert.

PATTERNS OF CONVECTIVE FLOW

Stein is a theorist who prefers sandals to shoes and a ponytail to a crew cut. For his doctoral thesis at Harvard, he calculated the acoustic noise the sun's convection zone generates, a topic we encounter in a later chapter. After a few years of teaching undergrads at Brandeis University, he moved to Michigan State University as an associate professor of physics. In 1983 he attended a conference in Copenhagen where he met Ake Nordlund, a young Danish theorist. The two men struck up a friendship and formed a collaboration that has changed both their lives.

In the early 1980s Nordlund was making numerical simulations of solar convection just under the solar surface. Convective cells can radiate much of their energy in these layers, and Nordlund's treatment of radiative losses was the best to date. He was successful in reproducing the appearance of the granules that lie just under the solar surface. But he was forced to make restrictive assumptions to simplify his calculations, and he was limited in the amount of detail he could resolve in his computational box. Stein and Nordlund agreed to work together to improve the simulations. They were able to include better physics and as faster computers came along, to resolve more detail.

After their calculation runs for a sufficient time, the plasma in their box seems to settle into a definite pattern of motions. This is not to say that two snapshots taken, say, a simulated month apart would look identical. The details of the pattern change constantly, but the same general features remain. These are the most robust results that have come out of these simulations and they give us the best picture so far of how convection in the sun behaves.

Figure 4.6 shows some of their results. The different columns show the temperature, the vertical velocity, the loga-

log Density	Temperature	Vz	Vh

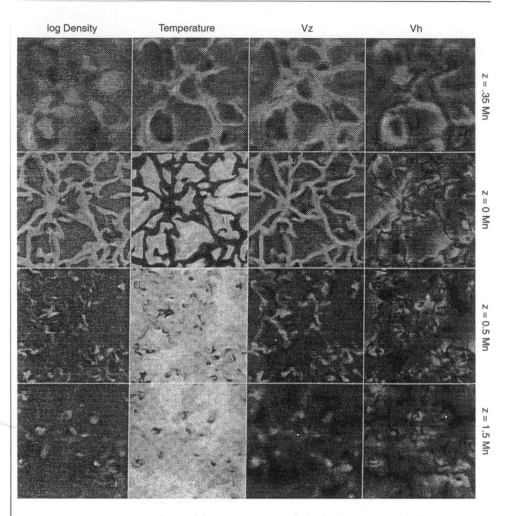

Fig. 4.6 Images of the plasma variables in granulation, from the calculations of Stein and Nordlund. The columns, left to right, show plasma density, plasma temperature, vertical velocity, and horizontal velocity. The rows are labeled with the depth, z, below the surface, in millions of meters. (Courtesy of Springer Verlag)

rithm of the pressure, and the helicity (twist) of motions, at several depths in their computational box. Notice that the layer they consider is only 2 megameters (2,000 km) deep, as compared to the 200,000 km depth of the real solar convection zone!

First, look at the tile-like pattern of convective cells in the second row. These computed cells match the observed properties of solar granules reasonably well: their brightness, sizes, shapes, speeds, and lifetimes. Moreover, several groups (for example, Fausto Cattaneo and Andrea Malagoni at the University of Chicago) are finding similar results. This agreement between independent groups and with observations, gives us some confidence in the other results of the calculations.

In the Stein-Nordlund simulations, the granules rise, hot and bright, at their centers. At the top of their rise they are free to radiate much of their load of heat to outer space. So they cool, become denser, turn over, and flow down at their edges. Just below the actual surface (the row labeled $z = 0$ in Figure 4.6) the downflows merge into fast, narrow, twisting funnels that plunge deep into the box and indeed out the bottom of the box. Thus the convective flow is *highly asymmetric*. Warm plasma rises gently in broad diverging upwellings, while cool plasma plunges in narrow funnels. The funnels themselves merge at greater depths, to form just a few powerful downflows. This picture was quite unexpected and contradicts the old pattern that was assumed in the mixing length theory.

In the mixing length theory, the flows are cellular at all depths, and the size of the cells at each depth is proportional to the local pressure scale height. As the large cells decay at the top of their rise, they generate smaller cells, which in turn create even smaller cells, and so on. Energy and momentum flow from large to small sizes.

In the new picture of Stein and Nordlund, small cells near the top of the box cool by radiating to outer space. Their cool, dense plasma descends in fast, deep funnels that help to turn over the larger, slower cells deeper down. Momentum flows from small to large sizes, the *reverse* of the mixing length description. At first glance this process looks like "the tail wagging the dog." But a careful look at the funnels shows that they contain most of the kinetic energy of the convection, while the largest cells contain most of the excess heat.

DEEP CONVECTION

Does this pattern of broad, hot upflow and narrow, fast, twisting downflow persist right down to the bottom of the convection zone? As we mentioned earlier, a simulation of the entire convection zone is beyond the power of present computers, so it is not possible to answer this question yet. But studies of convection in the deepest layers of the zone show effects similar to those near the surface.

Deep in the zone, there is no possibility of radiating energy directly to outer space. And so, at first sight, we might expect that the strong downflows in narrow, cool funnels, so characteristic of the upper layers of the zone, would not develop. However, the simulations of these layers do show a similar if weaker pattern, with broad upflows and narrow, twisting downflows.

Although the flow pattern in these deep layers differs from that in a mixing length theory, nevertheless there is a curious residual similarity. The sizes of the deep cells are proportional to the local pressure scale height, as in the mixing length theory. In fact, the mixing length theory works pretty well deep down in the zone, since one can now predict the correct size of the mixing length.

TURBULENCE

As we saw earlier in our first cruise up and down the zone, convective flows are highly turbulent when seen at higher magnification. In fact, every fluid encountered in life becomes turbulent when it flows fast enough. The water out of a tap, the wind around a house, the airflow around an airplane all exhibit turbulence—unruly, seemingly chaotic motions. Turbulence enhances the mixing of a fluid (as in the color blender in your local paint store) but it also bleeds energy from an otherwise smooth flow. Thus airplanes burn more fuel to feed the wasteful turbulence over their wings. So one of the principal tasks of an airplane designer is to limit or suppress turbulent flow around his machine in order to achieve high speeds at high efficiency.

Ludwig Prandtl, the inventor of the mixing length, also introduced the concept of the boundary layer, a thin region near an airplane's skin. Inside the layer, flow is turbulent; outside it, flow is nearly laminar. By conceptually separating the two regions in this way he was able to deal with the nasty phenomenon of turbulence in this special case.

The astrophysicist must also deal somehow with turbulence, no matter how difficult a task it may be. For one reason, the transfer of heat is much more efficient in turbulent convection. (After all, an egg cooks faster if the water boils, not merely simmers.) And the mixing of regions of different composition within a star is also enhanced by turbulence. So the astrophysicist must try to incorporate turbulence in his descriptions, but just how is a matter of controversy.

As we pointed out before, the hydrodynamic equations that govern convection are nonlinear, which makes them very difficult to solve. In turbulent flow, conditions change even more rapidly and on very small spatial scales, which further complicates any attempt to describe the phenomenon. Werner Heisenberg, one of the pioneers of quantum mechanics was thoroughly frustrated in his attempts to master turbulence. A story is told that as he lay dying, he said he would ask God two questions: why did He create relativity and why turbulence? God, he thought, might have a reasonable answer for the first question.

One of the first to grapple with the phenomenon was Osborne Reynolds, an English contemporary of Rayleigh, who, in 1912, devised a criterion for the onset of turbulence in a smooth flow. Three factors combine to decide when this will occur. First there is the viscosity v of the fluid, then the speed of the flow V, and finally a characteristic length L over which the speed is roughly constant. These quantities form a dimensionless ratio, the Reynolds number

$$R_e = VL/v.$$

One can demonstrate this relationship by stirring tea in a cup. The characteristic length L is the width of the cup. When stirred very slowly V in the formula is small), the flow is smooth or laminar. When stirred vigorously, the flow becomes turbulent. And if honey is substituted for the tea in the cup (honey's viscosity is much greater than water's), no amount of stirring will create any turbulence.

Experiments with different fluids showed that turbulence appears in a smooth flow at a Reynolds number of about a thousand. *Estimates of R in the solar convection zone lie around 10^{12}!* This very high number implies that even the tiniest volumes will contain turbulent motions.

Another milestone in the study of turbulence was reached in 1921 by Geoffry Ingram Taylor, the distinguished English hydrodynamicist. He introduced the idea of describing turbulent motions *statistically*. With such complicated motions in a fluid, it makes little sense to try to follow every little vortex at every moment. All the physicist can hope for is some insight into *time-averaged* quantities.

Taylor introduced two important averages. The first (the so-called correlation function) measures the degree to which the velocity at two points in the fluid fluctuates in step. In stop-and-go traffic the speed of a car affects the speed of the car behind it. In the same way, the speeds of different blobs in the plasma can be correlated. The speed at a particular point, at this moment, affects the speed at a different point, at a later moment. Taylor introduced a correlation function, which is the average, over a long time, of the product of the speeds at the two different points.

Taylor also introduced the so-called "power spectrum" which describes how, over the long run, kinetic energy is distributed among the different sizes of turbulent eddies. Both of these averages are useful only when the turbulence is so violent that it looks the same everywhere, in all directions (i.e., homogeneous and isotropic).

Andrei Nikolaevich Kolmogorov took the next important step in 1941. He was a Russian mathematician on the faculty of Moscow University. Early in his career he made a name for himself by placing the slippery concept of probability on a firm mathematical basis, that of set theory. He continued to make many important contributions to Soviet science and mathematics and was eventually awarded the coveted Stalin Prize. An expert in probability and equipped with excellent physical intuition, he was drawn to the challenging field of turbulence.

He applied the fundamental principle of "similarity," well known in gas dynamics, to the complicated motions in homogeneous isotropic turbulence. In essence he postulated that the flows in turbulence *look the same at any magnification* if you change the time and distance scales appropriately.

Suppose you made a movie of a large volume of turbulent fluid. Then suppose you zoomed in on a small piece of this volume and made another movie. If you played the second movie at a slower speed and with a slightly blurry focus, it would look much like the first movie. At every place in the fluid you would find the same distribution of small and large eddies, all interacting with each other.

In the interaction, energy flows in one direction, so to speak: from large eddies to small ones. In convection, buoyancy propels the largest eddies, which then energize the smaller eddies around them, and these in turn drive still smaller ones. One can imagine arranging all the eddies in the fluid in a chain of decreasing sizes, with a flux of kinetic energy flowing right down the chain to the molecular level where it turns into heat.

Kolmogorov realized that a kind of equilibrium must prevail among the eddies of medium size, between the biggest ones (which drive the whole system) and the smallest ones (where kinetic energy dissipates as heat). In this intermediate range Kolmogorov could predict the power spectrum, that is, the distribution of kinetic energy among eddies of different sizes. His famous result states that the kinetic energy E of an eddy increases with the 5/3 power of its diameter d:

$$E = d^{5/3}.$$

The Kolmogorov spectrum has been confirmed experimentally in the laboratory and is one of the few tools that modern simulations of convection can incorporate. However, observations show that turbulence in the sun and stars is neither homogeneous nor isotropic and is not even statistically the same from one moment to another. So the Kolmogorov spectrum is really not valid in the very places one would like to use it.

CHAOS, FRACTALS, AND TURBULENCE

Even so, great a physicist as Heisenberg was frustrated in his attempts to deal with the phenomenon of turbulence. The core of the difficulty lies in the fact that turbulence is neither completely random nor completely orderly. Moreover, it involves a continuous distribution of eddy sizes, all interacting with one another.

In recent years, a clearer understanding of turbulent flows has emerged with the development of the new science of chaos. Turbulence does seem to have a subtle orderliness after all, if one knows where to look. The place to look is the "strange attractor."

To understand this concept, imagine that you have isolated a tiny cube of plasma in the sun's turbulent convection zone. You could characterize the gross state of the plasma in the cube at any instant by measuring five average properties: the pressure, the density, and the three components of the velocity vector.

To record the behavior of the plasma you would want to plot these five quantities, at each instant of time, in a graph. A truly elegant graph would be one with five orthogonal coordinate axes, one for each quantity. Then you would need to plot only one point to represent the present state of the plasma. Such a graph is in fact a five-dimensional space (called "phase space") with a set of five coordinate axes. It may be difficult to visualize, but not impossible to

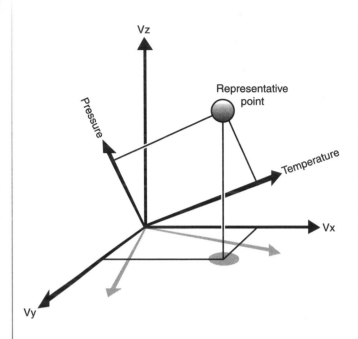

Fig. 4.7 Five-dimensional phase space. Each plasma variable (temperature, pressure, and the three velocity components) is plotted along one of five axes to fix a single point that represents the present state of the plasma blob.

conceive of, or to implement in a computer.

So, on each axis you mark off one of the five quantities and plot a single point in phase space that represents the instantaneous state of the plasma. (Figure 4.7 may help you to visualize this scheme.) As the plasma in the cube moves around turbulently, its five quantities change. You plot a new point for each moment. What is the result? Although in principle the point could leap all over the phase space in a completely random fashion, it doesn't. Instead, the representative point in phase space moves smoothly and traces out a complicated coiled curve (see Figure 4.8).

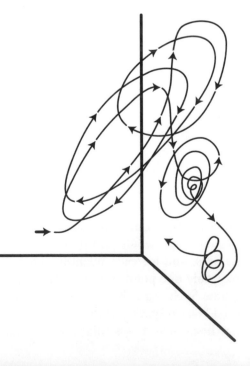

Fig. 4.8 As a plasma blob evolves in space and time, its representative point in phase space traces a complicated path, limited by a "strange attractor."

Moreover, the curve, which represents the time-history of the plasma, coils upon itself *without ever crossing itself.* The developing curve may approach an older coil infinitely closely, but like a long snake it never intersects itself.

In this sense the turbulent behavior of the plasma has a kind of order. Although the state of the plasma never exactly repeats (for then the curve would cross itself), it may come infinitely close to a former state. And the range of variation of its five properties is fairly limited: the pressure and density never go off to infinity, for example. Theorists of chaos refer to such a physical system as having a "strange attractor" which confines the state of the plasma to relatively limited values. Different kinds of turbulent flow systems have different kinds of strange attractors. Experimenters who study chaos often use turbulence as a practical model of a chaotic system.

Suppose you followed the cube of plasma for a very long time, plotting its state at each instant in the phase space. Then you zoom in on some region of the coiled curve in the phase space. You would see that indeed the coils are very close, without touching. Now if you magnify this region in phase space still further, you uncover still more coils, which were unresolvable before. And they do not touch either! You can keep increasing the magnification indefinitely, and find more and more densely nested coils in the same region.

You have discovered that the curve that represents the time-history of your sample of plasma is a *fractal.* A fractal is an object whose parts resemble the whole, and whose parts of parts resemble the whole, and so on to the smallest possible subdivisions. We can say that such an object is *self-similar*—it looks the same at all scales.

Fractals have received an enormous amount of public and scientific attention since their description in the 1960s by Benôit Mandelbrot, a Polish-born French mathematician. Mandelbrot began by studying similarities in the short- and long-term fluctuations of the stock market. He moved on to fluctuations in telephone conversations, the irregularities of the coast of England, and the swirls in turbulent flow. Eventually he created a unified mathematical description of fractal objects that include such diverse objects as clouds, mountain terrain, and ferns.

So in summary, perhaps the best shorthand way of describing a highly turbulent fluid, like the sun's convection zone, is to say it has a fractal, or self-similar structure. Kolmogorov understood this intuitively when he applied the concept of self-similarity in order to obtain a useful, if limited, description of turbulent flow.

Here in the convection zone we have a good example of the way astrophysicists' descriptions become more abstract and more elegant as they try to deal with Nature. We started with hot blobs rising and falling, a simple picture anybody can visualize, and we end up at the frontiers of chaos theory. Elegance alone is not the ultimate goal, however, but rather, predictive power. We shall see this same pattern of constant striving for deeper, more fundamental descriptions reappear over and over in our continuing journey.

Chapter 5

LOOKING INSIDE FROM THE OUTSIDE

I n 1980 two French astronomers spent three months at the U.S. South Polar Station in Antarctica to observe the sun with a specialized instrument. Their stay was anything but comfortable—an experience neither wanted to repeat—but they were rewarded with a great prize. They had a clear look into the interior of the sun.

As recently as 1970 the solar interior was an unexplored country. The best astronomers could do was to build theoretical models of the interior, using the most plausible assumptions and a very few constraining observations. We've been reviewing some of their models in the past chapters. We've seen how a "standard" model of the sun was devised, and how the first steps have been taken to predict plasma motions in the convection zone.

However, the proof of the pudding is in the eating, as the old saying goes. Theoretical models are all very well and good, but without observational checks, they are nothing more than elegant sand castles.

Since the mid-1970s, the situation has changed for the better. Astronomers now have a new tool with which to investigate the sun's interior. It's called *helioseismology*, the study of the sun's vibrations. Just as seismologists use the waves caused by earthquakes to probe the earth's interior, so astronomers can use solar vibrations to look into the sun's interior. For the first time, they can see inside a star and determine its properties, not from theory alone but with highly precise measurements. In this chapter we'll see how this new science originated, what its physical principles are, and how some of its results have upset long-standing views.

We begin with an old story: a scientist is curious and invents a new way of looking at the world.

Oscillations of the Sun

Professor Robert Leighton was generally regarded as one of the best physics teachers at Cal Tech. With an easygoing manner and a fund of fresh ideas he attracted a circle of talented young graduate students to work with him. His main interest was infrared instrumentation but in the late 1950s he turned his attention to the sun. He realized that he could modify an old instrument at the Mt. Wilson Observatory (a spectroheliograph) to explore the motions and magnetic fields at the solar surface. Together with his students, Robert Noyes, George Simon, Neil Sheeley, and Alan Title, he began to produce photographs of the sun's surface, in which rising or falling regions appeared as light or dark patches.

Very soon he discovered global patterns. Figure 4.2 shows one of these. The surface of the sun is covered with irregular cells in which the plasma flows horizontally at speeds around 300 m/s to the borders of the cell. This is the "supergranulation," a word that Leighton coined. The cells are about 30,000 km across and persist for around a day. We encountered these large convective cells in the last chapter.

Even more important was Leighton's discovery of patches that oscillate vertically with speeds of a few hundred meters per second and periods around five minutes. These patches were smaller than the supergranulation and larger than the ordinary granulation (Figure 5.1).

Fig. 5.1 Patches of the solar surface oscillating with a period of about 5 minutes. This image is a typical snapshot, showing the rising and falling patches. Also see color supplement. (Courtesy of SOHO/ SOI/MDI consortium)

At first astronomers thought these patches were nothing more than the ripples that rising granules would make as they reached the surface. But in 1966, Pierre Mein in France and Edward Frazier in the U.S. introduced a new method for analyzing the oscillations. They produced a so-called "diagnostic diagram" which hinted that the patches were not some local effect of granules but evidence of a global phenomenon.

Two groups of theorists picked up the hint and suggested an explanation that turned out to be correct. Roger Ulrich at UCLA and independently Robert Stein (at Michigan State University) and John Leibacher (at Lockheed Palo Alto Research Laboratory) proposed that the surface oscillations arise from standing sound waves in the convection zone.

In essence, they said that sound waves with a broad spectrum of frequencies are generated as by-products of the convective motions, just as boiling water makes plopping noises. Waves within a narrower range of frequencies are trapped within the convection zone. These waves interfere with one another and set up a standing wave. The oscillating patches at the solar surface are therefore parts of a complicated interference pattern of many waves of slightly differing frequencies—a "beat" pattern. The sun's surface vibrates like the top of a drum.

A key prediction of this theory was that the frequencies and horizontal wavelengths of the trapped waves should form a family of distinct curves in the diagnostic diagram. In 1975 Franz Deubner, a cheerful astronomer from Freiburg, Germany, made observations of sufficient quality to resolve the curves (Figure 5.2 is a more recent diagnostic diagram). At that point, the standing wave proposal was confirmed and the science of helioseismology was born.

With the theory confirmed, solar astrono-

Fig. 5.2 A "diagnostic diagram," which arranges individual modes of oscillation according to their frequency (vertical axis) and horizontal wavelength (horizontal axis). All the modes with the same number of nodes along a solar radius n fall on the same discrete ridge. The ridge at lower right corresponds to n= 0, the ridge at the upper left to n = 40. Also see color supplement. (Courtesy of SOHO/ SOI/MDI consortium)

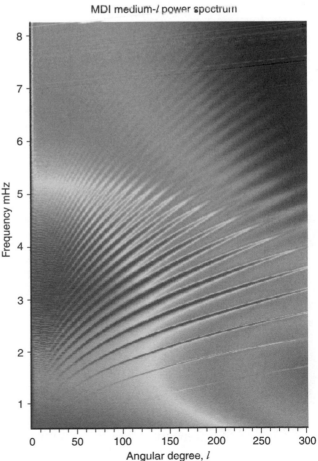

MDI medium-*l* power spectrum

mers were ready to move ahead. Rather than using observations to test the theory they now could use the theory to extract fundamental properties of the sun's interior from the observations. A key component of the theory, for example, is the temperature profile in the interior. Astronomers can now take a guess at a trial profile, use the theory to predict which waves get trapped, and calculate their frequencies. If the predicted frequencies match the observations, they have found the correct temperature profile; if not they have to try again.

THE PHYSICS OF SOLAR OSCILLATIONS

Convection is noisy—the sounds of a boiling kettle are familiar evidence of this. In the sun's convection zone, sound waves are generated by the fast turbulent downdrafts we encountered earlier, near the top of the zone. This sound is much too low in frequency for the human ear to detect (around one cycle every five minutes!) but if you analyzed it with the proper equipment you'd find that it is a form of very low-pitched noise, with many tones playing at once.

According to the oscillation theory, these sound waves propagate throughout the convection zone, like echoes in an empty ballroom. Most waves cannot escape from the zone. At the top of the zone, the steep gradient of plasma density acts like a mirror and reflects the waves, like ocean waves bouncing off a breakwater. Near the bottom of the zone the sound waves are gradually turned back or refracted toward the surface by the gradient of temperature. (See endnote 5.1 for details of refraction.)

Fig. 5.3 Sound waves are refracted or turned as they travel down into regions of higher temperature and faster sound speed.

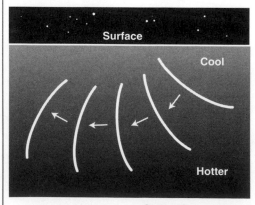

In this way, the wave moves in a series of arcs, turning around at some depth, returning to the surface, and reflecting there to continue on its way (see Figure 5.4). The wave is trapped between the surface and the depth at which it turns, so it travels within a specific spherical shell. A typical wave takes about five days to circle the sun.

Not all waves are trapped, however. Those with periods shorter than about three minutes can escape into the overlying atmosphere, where they eventually "break," like an ocean wave on a beach, and help to heat the atmosphere.

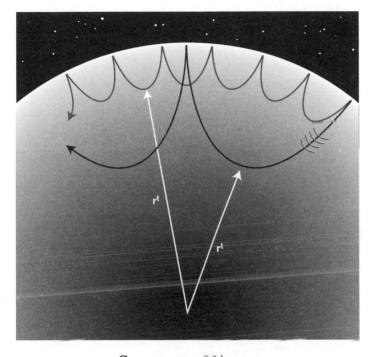

Fig. 5.4 Sound waves with short periods and short horizontal wavelengths do not penetrate as deeply into the sun as longer-period waves. (Courtesy of University of Arizona Press)

STANDING WAVES

This is still not the whole story, however. Each wave has a twin that travels around the sun in exactly the opposite direction. After circling the sun, one traveling clockwise, the other counterclockwise, these twins will meet and overlap. If their vibrations do not match exactly, point by point, they will destroy each other. If they do overlap exactly, however, they can set up a resonant standing wave (see endnote 5.2 on standing waves). Figure 5.6 illustrates how the vibrations of a particular pattern (or "mode") look in the three dimensions of the sun's sphere. Figure 5.7 shows a variety of possible patterns.

The sun turns out to be chock-full of standing waves. Figure 5.2, the "diagnostic diagram," shows the current state of the art. (To learn how many different kinds of waves [or "modes"] are packed into this diagram, see endnote 5.3.)

If you glance back at Figure 5.4 you will see that the fewer reflections a trapped wave makes at the solar surface (i.e., the longer is its horizontal wavelength) the deeper it penetrates the sun. So short-wavelength (high-frequency) waves travel through shallow layers of the convection zone and long wavelengths travel through deep layers. But as you can see even in this simplified figure, different waves can reflect near the same point on the surface. In the real sun,

Fig. 5.5 A standing wave on a violin string consists of twin waves traveling in opposite directions and overlapping. The three panels show the twin waves and their sum (in a double line) at three phases of the oscillation. In the top panel the waves overlap exactly, in the middle panel the waves have separated by a quarter of their wavelength, and in the bottom panel the waves cancel exactly. Notice how the nodes (the points where the amplitude of the oscillation is zero) remain fixed along the length of the string.

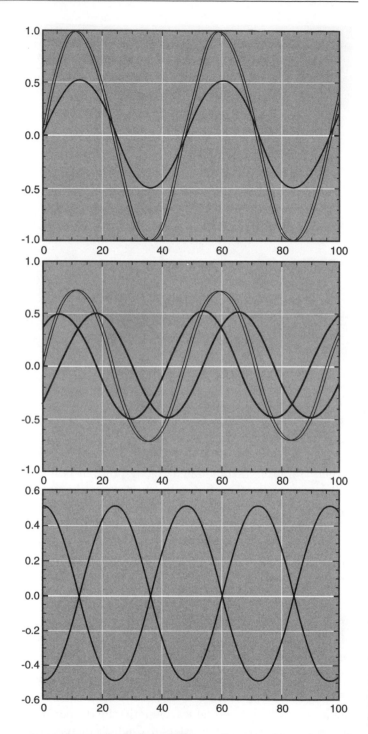

many waves with slightly different frequencies and traveling in slightly different directions at or below the surface can reflect near the same point.

Each mode is a vertical oscillation of only a few centime-

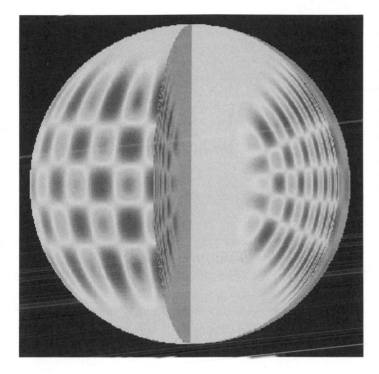

Fig. 5.6 An example of a three-dimensional standing wave pattern in the sun. Light grey regions are rising, dark grey regions are falling. Also see color supplement. (Courtesy of SOHO/SOI/MDI consortium. SOHO is a project of international cooperation between ESA and NASA.)

ters per second, which would be difficult to detect. However, the oscillations that are nearly in phase add up to produce an observable oscillation of several hundred meters per second. (One heckler may not be heard in a big hall,

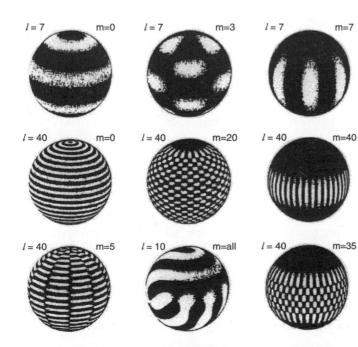

Fig. 5.7 A sample of the many standing wave patterns that overlap on the solar surface. They are distinguished by three "quantum numbers," n, l, and m. (Courtesy of University of Arizona Press)

but a crowd is sure to be.) What one sees at the surface, therefore, is a pattern of superposed waves with periods between three and about nineteen minutes. Figure 5.1 illustrates the surface oscillation patterns that several different modes produce. All of these and many, many more are present at the same time.

To keep things simple, I have avoided mentioning so far another distinct type of wave—the internal gravity wave. Sound waves are traveling *pressure* patterns, whereas gravity waves are traveling patterns of *buoyancy*. (The waves on the surface of a pond are examples of one type of gravity wave.) Theorists predict that solar gravity waves are firmly trapped below the convection zone. But even if they could somehow leak to the surface, they would still only produce oscillations of only a few millimeters per second, compared to hundreds of meters per second for sound waves. Since nobody has reported a confirmed gravity wave at the solar surface, we will ignore them and concentrate solely on the pressure waves, or *p*-waves.

OBSERVATIONAL TECHNIQUES

Astronomers have two basic methods for observing the solar oscillations. They can measure either the periodic *velocity* or the periodic *brightness* at each point of the surface. In addition, they can either make images of the oscillating patches as Leighton first did, or they can combine all the oscillations into one large signal.

To measure solar velocities, astronomers use the Doppler effect. A familiar example of the effect is the change in the sound of a train whistle as it approaches and passes. Similarly, the spectrum lines emitted by a rising or falling blob of plasma are shifted in frequency by amounts that depend on the blob's speed. By measuring this shift, one can determine the speed. (See endnote 5.4 about instruments.)

A helioseismologist's highest priority is to measure the oscillation frequencies as accurately as possible. By comparing the observed frequencies with those predicted from a temperature model of the sun, she can decide whether the model needs to be modified. Since the standard solar model is already fairly satisfactory in its predictions, frequency measurements must be precise to at least one part in ten thousand to be useful.

It's a fact of life that the longer one observes an oscillation, the more precisely one can determine its frequency. The basic process is to count the number of cycles in a fixed period of time. But unavoidably the count stops in the middle of some cycle and this residue is essentially the error in the count. This means that the larger the number of cycles counted, the smaller is the fractional error in the estimated frequency.

For some purposes, a precision of at least one part in ten thousand is required. But ten thousand cycles of five minutes each would require a continuous observation lasting thirty-five days. Unfortunately, even the most talented astronomers can't observe the sun at night. So there will always be day-night gaps in her counts of cycles, and these gaps will limit the accuracy of her measurement of a single frequency. To avoid this inconvenient feature of observing the sun, astronomers have devised a number of strategies.

THE ANTARCTIC

For six months out of every year the sun does not set at the South Pole. So if one can bear the cold, there's a chance to get six months of continuous oscillation data. In January 1980 Gerard Grec and Eric Fossat left their comfortable observatory in Nice, France, for the frigid Antarctic, in order to seize this scientific opportunity. They used a resonance absorption cell that measured the Doppler shift of a sodium spectrum line in the integrated light from the sun. This clever device detected the lowest meridional oscillations (with degree l = 0, 1, 2, 3) of the sun, with speeds as low as 2.5 cm/s—about as fast as an ant crawls. Figure 5.9 shows the spectrum of oscillations they obtained in a continuous five-day run. These low-degree modes probe the deep interior of the sun and, as we shall see, provided a critical test for the standard solar model.

Jack Harvey and Tom Duvall of the Kitt Peak National Observatory were inspired by this French breakthrough. In the austral summer of 1981, they traveled to the South Pole Station that the U.S. National Science Foundation operates. In collaboration with Martin Pomerance (from the Bartol Foundation) they measured the oscillations of brightness of a particular spectral line, over a two-dimensional grid on the solar disk. They had four uninterrupted periods of observation in three weeks, the longest of which lasted fifty hours. From their data they constructed a superb diagnostic diagram (similar to Figure 5.2) for degrees l between 4 and 140. At the low end of this range, individual modes were revealed for the first time as little dots.

Harvey and Duvall had to live in a room the size of a broom closet for two months in order to get their data. Seemingly gluttons for punishment, they returned for a second campaign in 1983 and their colleague, Stuart Jefferies, for a third in 1990. A recent campaign yielded continuous observations for sixty-five days, a rare event unlikely to be repeated soon.

These observers learned that the Pole's theoretical advantage of continuous sunlight is overrated. Ice fogs, storms, cirrus clouds, and occasional high winds conspire to limit the number of days of uninterrupted sunshine one can use. Moreover, the difficulties of maintaining equipment, to say nothing of living in so hostile an environment, are formidable. So observers have turned to other means.

THE NETWORKS

Fig. 5.8 A spectrum of the low-degree oscillations (l = 0, 1, 2, 3) of the solar surface, obtained by G. Grec and E. Fossat the the South Pole. (Courtesy of G. Grec, Solar Physics 82, 55, 1983)

Another way to follow the sun continuously is to establish a chain of identical observatories around the globe. Then the sun would never set on the astronomer's little empire. The University of Birmingham's team was the first to try out this idea. As early as 1979 they obtained oscillation spectra from two stations, using resonance absorption cells. They obtained 280 hours of data which resolved the modes l = 0, 1, 2, and 3 (Figure 5.8). In 1981, they tried again, this time with one station in the Canary Islands and another in Hawaii. Their observations—seventy days' worth—seemed to indicate that the sun's core is rotating two to nine times

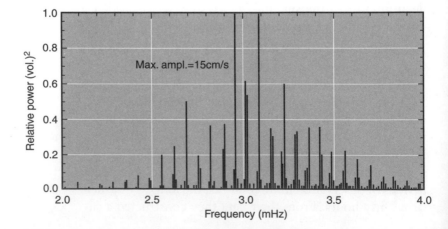

faster than the surface. This result was met with a lot of skepticism, however.

Encouraged by their success, the Birmingham group established BISON (the Birmingham Solar Oscillation Network) originally with two stations, now with six. Their effort was followed by the French, with the six-station IRIS (International Research on the Interior of the Sun); by the Taiwanese with the three-station TON (Taiwan Oscillation Network); and by an international group led by the U.S. with the six-station GONG (Global Oscillation Network Group).

The GONG project is a collaboration of over a hundred scientists, observers as well as some of the best theorists. The network consists of six stations deployed in California, Hawaii, Chile, Australia, India, and the Canary Islands and sees the sun about 90% of the time. Originally funded to operate for three years, it has been extended for a full solar cycle of eleven years because its results have been so exciting.

OBSERVATORIES IN SPACE

In some orbits, a satellite can see the sun continuously for months at a time. And there is no adverse weather to interrupt a long program. With these advantages, it wasn't long before a series of oscillation experiments was launched on satellites.

One of the simplest and most elegant was the Active Cavity Radiation Intensity Monitor (ACRIM). This very stable device, built by Robert Willson at the Jet Propulsion Laboratory, monitors the amount of solar energy (summed up over all wavelengths) the earth receives. Meteorologists interested in long-term variations of climate consider such data as vital. Martin Woodard and Hugh Hudson, two enterprising scientists at the University of California at San Diego, realized that the radiometer was sensitive enough and stable enough to detect variations of the sun's luminosity to a few parts per million. So they analyzed 137 days of data taken during 1980 and found the low-degree oscillations again. One of their results was that the modes have lifetimes of about a week.

At the time of writing, the Solar Oscillation and Heliospheric Observatory (SOHO) is in orbit around the L1 Langrangian point, a location in space where the gravitational attractions of the earth and sun just balance. From this vantage point SOHO can see the sun continuously for months at a time and is scheduled to stay aloft for at least three years.

Three oscillation experiments are on board. There is the Stanford University experiment, SOI (Solar Oscillation Instrument), which uses a narrow-band optical filter to produce a Doppler image of the sun every minute. GOLF (Global Oscillations of Low Frequency) is a resonance absorption cell built to observe the low-degree oscillations. VIRGO (Variability of Solar Irradiance and Global Oscillation) measures the lowest-degree brightness oscillations. These instruments are churning out data of unsurpassed quality that will occupy the modelers for years to come.

Figure 5.1 is a snapshot of the sun's five-minute velocity oscillations. All the modes pile up at the nodes in an almost random-looking wallpaper pattern. These are the kinds of images SOI and GONG generate, minute by minute for months and years. In three years of operation, SOI will churn out some *6,000 trillion bytes* of data! How can one possibly learn anything about the solar interior from such a morass?

THE TEMPERATURE PROFILE OF THE SUN

To extract useful information, we need a theory of how sound waves are trapped (such as Ulrich's original theory) and, secondly, a preliminary model of the temperature and density in the sun. For example, suppose we want to check the radial variation of temperature in the standard model of Bahcall and Ulrich. Then we have two choices. We can apply the sound wave theory to the model's temperature profile to predict the oscillation frequencies and their amplitudes and then compare the predictions and observations. Any discrepancies will require modifying either the temperature model or the theory or both. This is the so-called "forward" method of analyzing the oscillation data.

Figure 5.9 compares the predicted and observed frequencies of low-degree oscillations ($l = 0, 1, 2, 3$) and shows that the match was good to about 15 microhertz or 0.5% of the observed frequencies. This was a very respectable confirmation of a theory and a temperature model, one that any astrophysicist might envy. However, the uncertainty in the measurements was much smaller than the discrepancies.

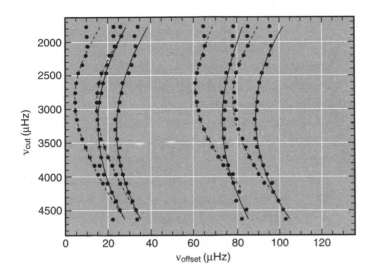

Fig. 5.9 When a spectrum like Figure 5.8 is rearranged, the patterns identify the different modes. A theoretical pattern computed from the standard solar model of Bahcall and Ulrich (1988) is compared with the observations (solid lines). The mismatch is small but indicated the model needed improvements. (Courtesy of J. Bahcall and R. Ulrich in Solar Interior and Atmosphere, A. N. Cox et al. eds., 1991, University of Arizona Press)

Clearly this 1988 "standard" model of Bahcall and Ulrich needed to be improved.

An alternative to the "forward" method is the "inverse" method, which relies on the fact that different modes are trapped in different layers of the sun. If one combines the observations of several waves that sample the same layer, one can get an empirical measure of the sound speed (and thus the temperature) in that layer. Then by examining deeper and deeper layers in this way, one can build up an empirical temperature profile.

Solar models come and go. As the precision and range of observed oscillation frequencies improved, the modelers constantly had to revise their work. In 1988 Tom Duvall and colleagues compiled a set of the best available frequencies, from observations they made at the South Pole and Ken Libbrecht made at the Big Bear Observatory. In 1991, J. Christensen-Dalsgaard and friends used these data to determine the radial temperature profile to nearly the center of the sun. Their result was so precise that it showed a distinct bump in the temperature gradient just at the base of the convection zone. The bump occurred at a fractional radius of 0.713 ± .003. Still better frequency data from the GONG project now place the base of the convection zone at 0.709.

Many other investigators have inverted seismic data. To obtain a seismic model of temperature and density throughout the interior, they often had to combine observations from different groups, made at different times with different types of instruments. Although these empirical models agreed very well with the best of the current theoretical models, there still remained small discrepancies, especially near the core, where the observations were sparse.

In a sense, measurements of the temperature in the core are the most interesting of all, because they bear on the critical neutrino deficit problem. A very small decrease in the central temperature could easily erase the deficit, so high precision is required. But the core is best sampled by the low l modes, and these are the most difficult to observe.

As we have seen, Gerard Grec and his French associates obtained the first low l frequency measurements at the South Pole in 1980. These were followed by a long series of observations by the Birmingham group, culminating in the BISON network. As new observations accumulated, estimates of the core temperature continued to fluctuate slightly. Part of the uncertainty could be attributed to the fact that both groups were using resonance cells, which do not usually resolve the spatial patterns of the oscillations.

Steve Tomczyk and his colleagues at the High Altitude Observatory therefore decided to build a new instrument, the low degree or LOWL, to correct this deficiency. In 1995, after six months of operation at the Mauna Loa Observatory in Hawaii, the instrument yielded precise frequencies for l = 0 to 80 in the range of 1.5 to 3.5 millihertz. The data were then inverted by Sarbani Basu (University of Aarhus, Denmark) and his associates and compared with a set of temperature models. Their models are nonstandard in having less helium in the core than the usual 28% of the mass. With this adjustment *the speed of sound derived from the observations agree with the model predictions within 0.15%, down to a fractional radius of 0.05!*

Their work, like several others before it, indicates that the solution to the neutrino problem cannot lie in some nonstandard solar model, but rather in elementary particle physics. It is fascinating to see how different subjects in physics connect.

The SOHO satellite is now churning out superb data for all the modes between l = 0 and 300. The low l data are especially valuable since they reveal conditions near the solar core. A large team of investigators has analyzed sixty days of continuous observations obtained during 1997. Figure 5.10 shows the differences they found between the empirical square of the sound speed (which is proportional to the temperature) and those predicted by a particular solar model. The maximum discrepancy (0.4%) appears just be-

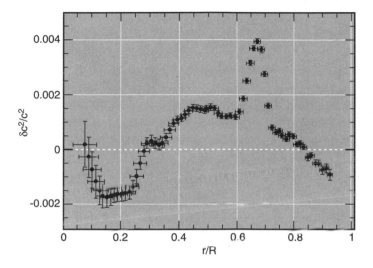

Fig. 5.10 A comparison of predicted and observed sound speeds along a solar radius. (The square of the speed at any depth is proportional to the plasma temperature at that depth.) The maximum discrepancy is 0.4% at three-quarters of the solar radius. (Courtesy of SOHO/SOI/MDI consortium)

low the base of the convection zone, at 0.67 R, and is probably due to a deficit of helium there. Also, notice the sharp decrease of sound speed relative to the solar model at the edge of the core, 0.25 R. This effect is unexplained at present but may arise from an overabundance of helium in the solar model.

Some Headlines from the Front

Helioseismology is booming. Every month some new discovery about the sun's structure and behavior is published in the journals. Here we can only sketch a few of these successes.

Solar astronomers have been debating for years about the amount of helium in the sun. Different methods and different data yielded values ranging between 20% and 40% of the sun's mass. It may seem surprising that the precise amount of the second most abundant element is so poorly known. Part of the problem has been that helium emits relatively weak spectrum lines in visible light and very strong lines that are difficult to interpret, in the extreme ultraviolet. What's more, direct sampling of the composition of the solar wind (see chapter 11) reveals real variations in helium abundance of at least a factor of two.

This uncertainty allowed astrophysicists to tweak solar

models to account for the observed deficit of solar neutrinos by lowering the accepted abundance of helium in the core. However, in order to match the frequencies of low-degree modes that we just discussed, Christensen-Dalsgaard and friends had to invoke an *increase* in abundance of helium in the core. This, they proposed, arises from gravitational settling of helium from the radiative zone, which predominates over diffusion of helium out of the core. Unfortunately the LOWL and SOHO data now imply that helium is *less* abundant in the core than the best theoretical models assume. The best estimate of the *average* abundance of helium in the sun now lies between 23.2% and 25.3% of the solar mass, lower than the standard model's 28.0%.

In 1985 Martin Woodard and Robert Noyes startled the scientific community by announcing that the frequencies of low-degree oscillations change in the eleven-year solar cycle. They had analyzed the brightness oscillation data from ACRIM, the luminosity monitor we mentioned earlier, for the interval between the cycle maximum (1980) and the minimum (1985). The low-degree frequencies had decreased during this time by 0.4 microhertz or one part in ten thousand. Since then the effect has been confirmed, not only by later ACRIM data but also independently by the Nice and Birmingham groups, with their ground-based resonance absorption cells. The frequencies rose again at the next cycle maximum in 1991. Some theorists attribute the swing in oscillation frequencies to a cyclic change in the temperature profile of the deep interior.

These changes in oscillation frequencies are only part of the story. In 1988, Robert Willson, the designer of ACRIM, and Hugh Hudson announced that the *luminosity* of the sun changes with the eleven-year cycle, by a mere 0.001%. The sun is brighter at the cycle's maximum despite the many dark sunspots that cover a portion of the disk at the maximum of the cycle. So the sun is a variable star, but at a level that is almost imperceptible.

All the results mentioned so far concern *global* properties of the sun—its temperature profile, helium abundance, and so on. More recently, astronomers have used the oscillations to derive the properties of *isolated* features located near the sun's surface. Among these are sunspots, the dark blemishes that contain strong magnetic fields. In 1988 Douglas Braun, Thomas Duvall, and Barry LaBonte developed a special method of analyzing the fate of traveling sound waves in and around sunspots. To their surprise they found that half the acoustic power that entered a spot is absorbed. The amount of power is too small to heat the dark spot, but the question remains as to what happens to the sonic power. In any case the absorption effect is likely to yield new information on the subsurface structure of spots.

In March 2000, Braun and his colleagues announced that they could actually detect sunspots on the *back side* of the sun, by analyzing solar oscillations!

In 1980, R. F. Howard and B. J. LaBonte discovered a kind of solar "jet stream." A narrow band, in both the northern and southern solar hemispheres, rotates about 2 m/s faster than the average speed. These bands originate at high solar latitudes and migrate toward the solar equator during the eleven-year solar cycle. In August 1997, the oscillation team for the SOHO satellite reported the detection of *six* bands, that lie at a depth between 2,000 and 9,000 km *below* the surface, and flow about 5 m/s faster than their surroundings. Presumably these bands are related to the pair at the surface. As yet no physical explanation has been offered for their existence, but as we shall see in chapter 13, they are probably features of the large-scale circulation that produces the sun's magnetic cycle.

ROTATION OF THE SUN'S INTERIOR

Some of the most interesting helioseismic results concern the rotation of the interior of the sun. Studies of solar rotation date back to the work of Richard Christopher Carrington in the 1850s.

Carrington was the son of a rich British brewer. His father intended that he should study theology at Trinity College, Cambridge, but he was drawn irresistably to astronomy. With his father's wealth he eventually built his own observatory and became a dedicated observer of the sun. For the eight years between 1853 and 1861 he faithfully recorded the positions and motions of sunspots. From his observations he concluded that the sun's surface doesn't rotate rigidly.

Many studies of the surface rotation have been made since Carrington, mainly with spectroscopic measurements of the Doppler effect. As with any topic about the sun, the more information one has, the more complex the phenomenon seems to become. Depending on how and what one measures, one finds slightly different latitude variations of the angular speed. In crude terms the solar equator rotates in twenty-five days while the regions near the poles take about thirty-five days. Ever since Carrington, astronomers have wondered how the interior of the sun rotates.

Theorists tell us the simplest pattern of internal rotation, other than a rigid-body rotation, would have the angular speed constant on cylindrical shells that are centered on the sun's axis (see Figure 5.11). To agree with the surface observations, the angular speed would have to decrease inward along a radius in the equatorial plane.

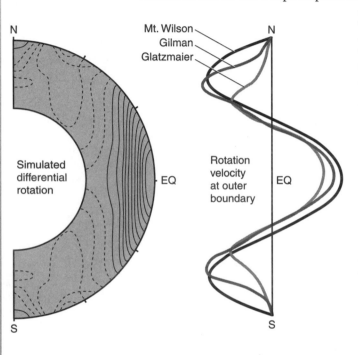

Fig. 5.11 Early computer models of solar rotation predicted that rotation speeds should be constant on cylindrical surfaces that are centered on the sun's rotation axis. These predictions turned out to be wrong. (Courtesy of University of Arizona Press)

Tom Duvall and Jack Harvey, those intrepid Antarctic astronomers, provided the first conclusive refutation of this simple picture in 1986. Using the observations they made at the Pole in 1981, they showed that the surface pattern of rotation persists all through the convection zone. In other words, the angular speed is not constant on *cylinders*, but along *radii* from the sun's center. Their conclusion was confirmed by a series of independent observers, including Tim Brown and Chris Morrow at the Sacramento Peak Observatory, Ken Libbrecht at the Big Bear Observatory, and Steve Tomcyk at the Mauna Loa Observatory. (See endnote 5.5 to learn how rotation is determined.)

In contrast to the convection zone, the radiative zone rotates *rigidly*. Figure 5.12 displays a recent empirical map of rotation speeds in the convection zone, extracted from the GONG observations. In the equatorial plane the speed rises to a maximum just below the surface and then declines inward to the base of the convection zone. At a latitude of 45

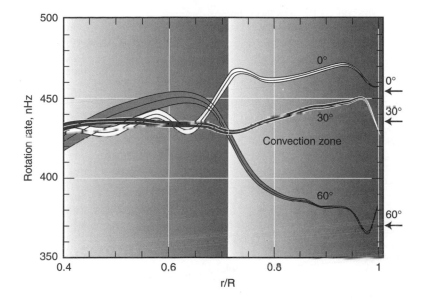

degrees, the rotation speed is nearly constant from the top to the base of the convection zone. Near the poles, the speed increases sharply from the surface to the base. In addition to this pattern, a fast polar "jet" appears at high latitude. Data obtained from the SOHO satellite since 1997 have confirmed these results in detail.

Fig. 5.12 A recent map of internal rotation speeds, obtained from observations from the SOHO satellite. (Courtesy of SOHO/ SOI/MDI consortium)

The radiative zone, below the convection zone, rotates rigidly, down to 0.4 R, at an angular speed intermediate between the highest and lowest speeds at the surface. This means that the convection zone must *rub* against the radiative zone, especially at low latitudes. In a narrow layer between them, only a few hundredths of a radius thick, strong shearing motions must exist. As we shall see in chapter 13, this shear layer is a prime suspect for the site of the generation of solar magnetic fields.

The rotation of the core, below a fractional radius of, say,

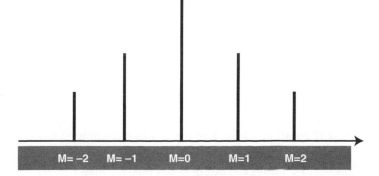

Fig. 5.13 Solar rotation splits the frequencies of individual l-modes into clusters or "multiplets," whose members correspond to different azimuthal modes, labeled by the quantum number m.

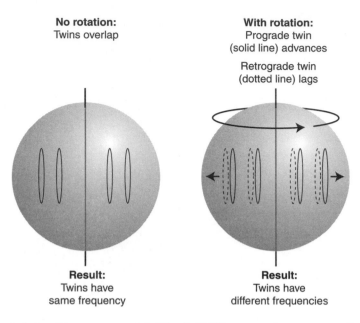

No rotation:
Twins overlap

Result:
Twins have
same frequency

With rotation:
Prograde twin
(solid line) advances

Retrograde twin
(dotted line) lags

Result:
Twins have
different frequencies

Fig. 5.14 The physical origin of frequency splitting. In principle, one can determine the frequency of a standing wave by counting the nodes of its constituent twin waves as they pass the central north-south meridian on the solar disk. Since the twin traveling in the direction of rotation is accelerated and the other twin is decelerated, the number of nodes counted for each twin in a fixed time (i.e., its frequency) will be different. In other words, the twins' frequencies have been modified (shifted or "split") by rotation.

0.4, is still controversial. The LOWL group has measured the pattern of rotation between 0.2 and 0.85 of the solar radius R. As of 1995, their best estimate of rotation frequency at $0.2R$ is 0.42 microhertz. For comparison the rotation period of the equator at the surface is 0.462 microhertz.

On the other hand, in 1996 Luigi Paterno combined early results from the IRIS network (l = 1 and 2 modes) with Ken Liebbrecht's 1988 Big Bear observations (l = 5 to 59) and claimed the region below $0.3R$ rotates 50% *faster* than the equator at the surface, that is, about 0.69 microhertz. And the data from the BISON network imply that the core rotates *slower* than the surface *polar* rate of 0.33 microhertz.

So there is still much work to be done to reach a consensus. However, the earlier claims of a core rotating two to nine times faster than the surface seem to be discredited.

An empirical map of the internal rotation of a star, like Figure 5.12, is a triumph of modern astrophysics. It could only be obtained with a combination of exquisitely precise measurements and a detailed theory.

YES, BUT WHAT DOES IT MEAN?

Now that solar astronomers have such a good map of the internal rotation of the sun, they'd like to understand the underlying physics that produces the unexpected pattern they see. Full understanding could be demonstrated with a numerical model, based on detailed physical theory, that would reproduce the derived map of Figure 5.12. Much progress has been made toward this goal but many important aspects are still unclear.

The basic problem is to understand why the internal rotation is *differential*, or why the different layers in the convection zone slide over each other at different speeds. The real complication is that the rotation speed is observed to vary in latitude as well as depth. This implies that some mechanism exists to transport angular momentum from high latitudes toward the equatorial plane.

The fact that the radiative zone rotates rigidly, while the convection zone rotates differentially suggests that convective motions are responsible somehow for the transport of momentum. But how?

Theorists face several obstacles in simulating solar rotation in a computer model. First, the equations that govern the coupling of rotation and convection are nonlinear and therefore difficult to solve, even numerically. Then the range of sizes of convective cells that should be included in a realistic model is too large to handle in even the most powerful computers. But the essential difficulty is conceptual: a lack of understanding of the behavior of convective cells and of the turbulence they produce.

As early as 1895, Osborne Reynolds pointed out that turbulence introduces stresses in a fluid, which act like friction between layers that slide over each other. This type of friction is like viscosity in that it tends to equalize the speed of sliding layers. Turbulence is much stronger than the ordinary kind of molecular viscosity and therefore must be included in any realistic calculation. But the magnitude of the turbulence depends on the mean flow and the mean flow depends on the magnitude of the friction. Nonlinearity again. But even worse, the turbulent forces are *anisotropic*. Since they arise from convective cells that rise radially and spread out laterally, the turbulent motions are different in different directions.

With such complications facing them, theorists are forced to take shortcuts. A common approach has been to include a reasonable range of convective cell sizes and to represent turbulence as a kind of microscopic viscosity.

Peter Gilman, an expert hydrodynamicist at the National

Center for Atmospheric Research in Boulder, Colorado, used such an approach for many years. He had access to the powerful supercomputers at the Center and was able to incorporate many subtle physical effects in his hydrodynamic codes. He assumed that convective "giant cells," much larger and deeper than the observed supergranules, drive the smaller cells. He also adopted a form of isotropic turbulent viscosity. With these assumptions and a huge time-dependent computer code he was able to reproduce the observed latitude variation of rotation at the surface, which was greeted as a substantial step forward.

However, his satisfaction was short-lived. As improved helioseismic maps of solar rotation appeared he realized that his models predicted the wrong pattern *inside* the sun: constant rotation speeds on cylindrical surfaces (Figure 5.11). Gary Glatzmaier, Gilman's colleague, found very similar results using his own independent computer model.

Several groups are trying to do better. Julian Elliot and Juri Toomre at the University of Colorado are using the fastest computers available to model rotation in a convecting, rotating shell. They incorporate isotropic turbulence in their simulations with a large range of convective sizes. To date, their model reproduces the solar rotation at high latitudes reasonably well, but near the equator they find, once again, cylindrical rotation, in conflict with the observations.

Bernard Durney, another seasoned contender in this competitive field, is presently at the University of Arizona. A self-confident scientist with superb physical intuition, he prefers to work alone and without recourse to the largest of computers. He is betting that the answer to the rotation problem lies in *anisotropic* turbulent eddies.

Durney has developed a theory of rotation that incorporates such eddies explicitly. He does this by choosing as free parameters the ratios of the eddy's three dimensions to the local mixing length. His theory does not *predict* these ratios, but once they are fixed, all else follows. He hasn't attempted yet to reproduce the helioseismic rotation maps, but is seeking instead to understand how the critical ratios depend on the mean speed of rotation and on the temperature gradient in the convection zone. He has used observations of the surface meridional flow (from the equator to the poles at about 10 m/s) to constrain these ratios and therefore the forces that produce them.

Günther Rüdiger is the latest in a long line of distinguished hydrodynamicists from the University of Potsdam. He and his colleagues have developed a very complicated description of anisotropic turbulence in terms of the mean rotation speed at each depth. By adjusting a few free parameters of the theory, they are able to predict a very good match to the observed pattern of rotation (see Figure 5.15). But this is still not the ultimate theory.

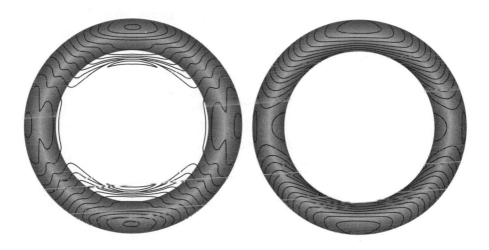

The experience of these experts reveals that some special piece of physics is still lacking in even the most sophisticated models and that it may be too soon to expect a realistic prediction of the observed rotation.

Fig. 5.15 A comparison of empirical (left) and theoretical (right) maps of internal solar rotation. (Courtesy of G. Rudiger)

The Origin of the Oscillations

So far we've just accepted the fact that the sun oscillates but it is certainly reasonable to ask what causes this behavior. No final answer can be given just now, but a very recent observational study offers a clue.

We hinted at the answer in the last chapter when we described the simulations of Stein and Nordlund. They found that convection cells merge into fast, narrow downflows, between slow, broad upflows. The downflows were thought to be turbulent and therefore good candidates for producing acoustic noise. That guess turns out to be correct.

Thomas Rimmele and his associates have observed acoustic "pulses" in the lanes between granules. When a lane becomes very dark very suddenly, a pulse of sound energy (literally a thunderclap!) is launched upward into the atmosphere. A short distance from the location where the darkening occurs, the photospheric plasma *begins to oscillate with a five-minute period.*

Rimmele and company suggest that the darkening heralds a dramatic sudden cooling of plasma in the lane. Two things follow: the cool plasma sinks ("like a bowling ball in a swimming pool") and a pulse of sound is sent upward. The turbulent sinking plasma creates a wide spectrum of sound waves that propagate in all directions. Most of them decay but they set the local atmosphere "ringing" at its natural period of about five minutes.

But there is nothing new under the sun. Rimmele and friends pointed out that the behavior they observed was predicted in 1906 by Sir Horace Lamb, one of the foremost hydrodynamicists of his time. He wrote a classic book that lays out the beautiful mathematics of the subject. But as one student remarked, "You could read the whole book and never realize that water is wet."

In summary, helioseismology is one of the hottest subjects in astrophysics today. With powerful instruments in space and on earth pouring out torrents of data, and with literally hundreds of bright scientists worldwide cooperating to interpret the data, we are discovering just how cleverly our old sun is put together.

Chapter 6

THE HOT ATMOSPHERE

t last we reach the surface of the sun, where sunlight finally escapes into deep space. Astronomers name this place the photosphere, after "photos," the Greek word for light. It's well named, since it is a dazzling, intensely brilliant inferno. At least 99% of all the energy that was produced in the core escapes as radiation from this level. All around us lie the granules, each as large as Brazil, packed edge to edge as far as the eye can see.

But the sun doesn't end here by any means! If we look up we see that a vast tenuous atmosphere rises far above us. Like sailors on an ocean we bob up and down on the dense plasma beneath us and gaze into the thin plasma above. Only in the sense that the plasma density falls off very rapidly above us does the sun have a real "surface."

This enormous atmosphere extends many solar radii into space and parts of it escape the sun entirely to stream all through the solar system, where it is known as the solar "wind."

A century ago the biggest puzzle in solar physics was why such a huge atmosphere exists at all. The corona had been seen repeatedly at total solar eclipses but until 1940 astronomers had no idea its temperature soars to 2 million kelvin. The reason for the sun's extensive atmosphere then became clearer: it exists because it is hot. A form of energy that we have not considered so far (nonradiative energy) heats the atmosphere and causes the plasma to expand upward, like the air above a stove. A constant flow of this extra energy maintains the high temperature so that the atmosphere glows in ultraviolet light and in X-rays. We know this extra energy cannot be thermal—neither radiation nor convected heat—because heat would not flow spontaneously from the relatively cool photosphere to the hotter atmosphere. But the actual mechanisms that heat the atmosphere are still

Fig. 6.1 A view of the
solar chromosphere in
the light of hydrogen H
alpha. In the fore-
ground lies an active
region. The edge of the
solar disk is at the top
right corner. All the
filamentary threads
outline magnetic fields.
(Courtesy of R. B.
Dunn)

controversial. (We'll discuss them at length in chapter 11.)

Even more puzzling to nineteenth-century astronomers was
the variety of structures they could see in the sun's atmosphere (Figure 6.1). The atmosphere is nothing like a
smooth continuous "fog" of plasma. Instead, a jumble of
strange objects greets the eye. Arches of hot plasma are
everywhere, in all sizes and intensities. Thin spikes of plasma
(the "spicules") shoot upward at blazing speeds. Active regions, huge arcades of hot plasma, sputter and fume dramatically up to heights of twenty thousand kilometers and
more. Embedded low in such regions are dark sunspots,
with bright hot plumes of plasma rising above them. Im-
mense fan-shaped cathedrals of tenuous plasma, the coronal streamers, loom high above the smaller features.

Only since the early 1900s have astronomers gained any
insight into the origin of all this complexity. The key to it
all is the realization that the sun is a magnetic star. Its sur-

face is covered with arches or loops of magnetic flux that
pop up from below, that migrate, interconnect, expand upward, and eventually dissipate. The magnetism interacts
with convective motions to produce violent, sometimes
explosive behavior. In addition, all this magnetic structure
changes systematically in stately cycles of eleven and twenty-
two years.

In the continuation of our journey, up through the solar
atmosphere, we shall follow these two themes: the flow of
nonradiative energy from the interior and the hot atmosphere it produces; and secondly the origin and behavior of
the magnetic field that threads through the atmosphere. As
we shall see, the two themes are intimately connected.

We begin by recalling how astronomers have used the solar spectrum to derive the temperature throughout the atmosphere. Along the way we'll need to discuss how atoms
emit and absorb light and how they behave in hot plasma.
Once we have a map of the temperature and density in a
simplified "smooth" atmosphere, we can begin to ask questions about how much power it takes to maintain them and
where that energy comes from.

THE PHOTOSPHERE AND ITS SPECTRUM

Nearly everything we know about the solar atmosphere has
been extracted with great labor from the analysis of its spectrum. And the spectrum was revealed only slowly, as astronomers gained the necessary means—the visible part
first, then the ultraviolet and infrared, and finally the X-ray
parts. As we shall learn, these different ranges of the spectrum originate primarily from different parts of the atmosphere, with the shorter wavelengths emitted from the hotter regions.

William Herschel, the discoverer of Uranus, was the first
to get a glimmer of how hot the photosphere is. In 1800 he
made measurements of the solar constant, the amount of
energy falling on a square meter on earth in a second, by
timing the rise of temperature of a bowl of water. When he
extrapolated his constant back to the sun he obtained a staggering figure, that convinced him the solar surface was indeed extremely hot. But the concept of a black body was
unknown in his time and so he was unable to obtain a definite temperature.

Technology has improved enormously since the days of

Herschel. We now know the earth receives 1.368 kilowatts per square meter, and so the photosphere must emit 63,000 kilowatts per square meter. A black body at a temperature of 5,770 kelvin would radiate this much power, so we can say the photosphere has an "effective" temperature of 5,770 kelvin.

We can check whether the photosphere really does radiate like a black body by looking at its spectrum. Isaac Newton was the first (1670) to spread sunlight into its constituent colors, using a glass prism. Herschel, once again the pioneer, measured the relative brightness of the different colors with his trusty thermometer, and discovered the infrared (or "heat") end of the spectrum. Figure 6.2 compares a

Fig. 6.2 A comparison of black body curves and the continuous spectrum of the photosphere. (Courtesy of D. Labs and H. Neckel)

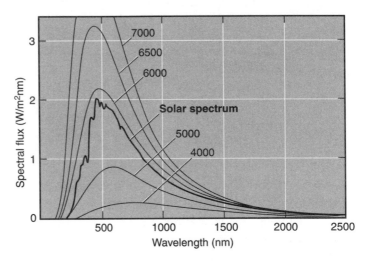

modern measurement of the photosphere's spectrum to a set of black body curves. As you can see, the fit is quite good for a temperature around 6,000 kelvin. The intensity peaks in the green part of the spectrum, where, not surprisingly, our eyes are most sensitive and where chlorophyll works well.

The visible part of the spectrum (from 360 to about 800 nm) is riddled with thousands of so-called absorption lines. Figure 1.4 shows a few of these. They are narrow bands (typically 0.01 nm) where the light intensity is somewhat reduced. William Hyde Wollaston, an English physicist, discovered them around 1814, and Joseph von Fraunhofer, a German instrument maker, measured the wavelengths of about five hundred of them soon afterward.

Forty-five years passed. At the University of Heidelberg, Gustav Kirchhoff, one of the giants of German experimental physics, was learning how solids and gases emit light. Kirchhoff made many significant contributions to science, especially in electromagnetism and in electrical circuit theory, but the most important was his discovery of several essential laws that govern radiation.

First, a heated solid emits a smooth, continuous spectrum, a "continuum." The glow of the filament in a light bulb is a good example.

Secondly, good absorbers of light are also good emitters. Think of the bricks in a furnace. If they are to maintain a stable temperature, they must radiate as much energy as they receive.

Third (and this is the most abstract but most important of Kirchhoff's discoveries), the emitting and absorbing powers of all substances in thermal equilibrium are related by the same universal function of wavelength and temperature. Max Planck would later prove that this universal function is in fact his famous expression for the intensity of black body radiation.

Unlike solids, gases emit narrow spectral lines. Moreover, each gas emits its own distinctive set of wavelengths. For example, sodium vapor emits two powerful lines in the yellow part of the spectrum, which are now used to illuminate highways.

When Kirchhoff passed the continuous light of a hot solid through a sodium vapor the two narrow lines appeared in absorption. Eureka! Kirchhoff was struck by two important ideas. First, every element has its own individual spectrum which is as distinctive as a fingerprint. That meant he could identify terrestrial elements by their spectra alone, not relying on their chemical properties. He went on, with Robert Bunsen (of the famous Bunsen burner), to identify dozens of terrestrial spectra, and discovered two unknown elements (cesium and rubidium) by their unique spectra.

Secondly, Kirchhoff concluded that a cool gas overlying a hot solid could explain the Fraunhofer spectra of the sun. He pictured the photosphere as a glowing solid or liquid surface (this was 1860, remember) lying under a cooler gaseous atmosphere. Just as in his laboratory, the cool gas absorbed the bright light of the solid and produced absorption lines. Now he could go on to identify terrestrial elements in the solar atmosphere!

Kirchhoff's work laid the foundations of spectroscopy, radiation theory, and, indirectly, astrophysics—a record any physicist can envy. Unfortunately, Kirchhoff's image of a hot solid surface and a cooler atmosphere misled three generations of astronomers.

Although Kirchhoff identified many terrestrial elements in the sun, he didn't discover any new ones. That privilege was reserved for two astronomers, Pierre Janssen and Sir Norman Lockyer. At the total solar eclipse of 1868, they observed a strong yellow emission line in a luminous cloud above the moon's edge. At first they thought it was a blend of the well-known pair of lines of sodium but the wavelength was a bit too short. Lockyer tried hard to find an element that emitted this pair of lines and finally concluded that Janssen had discovered a new element and named it helium, from the Greek word "helios," the sun. Helium is a minor constituent of the earth's atmosphere (only 1 part in 200,000), but as a "noble" gas it forms few chemical compounds and therefore is difficult to detect.

That same year, Anders Ångström, a Swedish physicist, published his wavelength measurements of over a thousand Fraunhofer lines, including the hydrogen lines for the first time. Ångström introduced the metric system for wavelengths. The standard unit, the Ångström, equal to one-hundred millionth of a centimeter, was named in his honor. (But fashions change and now the scientific world prefers the nanometer.)

Henry Augustus Rowland, an experimental physicist at Johns Hopkins University, extended the work of Kirchhoff and Bunsen. The critical piece of a modern spectrograph is its diffraction grating, which spreads light into its component wavelengths far more effectively than the glass prism that Kirchhoff used. Rowland produced the best gratings of his time, and used them to resolve and catalog more than twenty thousand Fraunhofer lines. By 1897 he had identified thirty-nine elements in the solar spectrum.

Much further work was needed before the strengths of these absorption lines could be interpreted to yield a quantitative chemical analysis of the photosphere, but eventually the relative abundances of the elements was determined (see Table 1.2). Hydrogen and helium turned out to be the two most abundant elements.

Kirchhoff's picture of the photosphere as a hot solid over-lain by a cooler atmosphere was replaced in the early 1900s by the work of two distinguished English astrophysicists, Arthur Eddington and Edward Milne. By then the sun was recognized as a hot ball of hydrogen, with a few traces of terrestrial elements. A more realistic picture of the photosphere was needed.

Milne and Eddington recognized that we see different layers in the photosphere when we shift from the wavelength of an absorption line to a nearby wavelength. The reason the absorption line is there at all is that some kind of atom in the atmosphere, like sodium or magnesium, emits and absorbs light of that wavelength preferentially. (This, remember, was how Kirchhoff was able to identify the responsible atoms.) That means the photosphere is especially *opaque* at that special wavelength. We cannot see very far into the photosphere at the wavelength of an absorption line and therefore the light that we receive must have been emitted from a relatively shallow layer. It's like looking through a dense fog—you can only see objects ahead a certain distance, depending on how thick the fog is.

In contrast when we look at the photosphere at wavelengths outside an absorption line, in the "continuum," the gas is less opaque and so we see deeper, to where the gas is hotter and therefore brighter. Unlike Kirchhoff's picture of a solid surface illuminating a gaseous atmosphere, the photosphere of Eddington and Milne is a continuous gaseous atmosphere in which the temperature decreases smoothly outward.

In their models of the photosphere, the central intensities of the lines map the outward decrease of black body temperature (see Figure 6.3). We now know the temperature

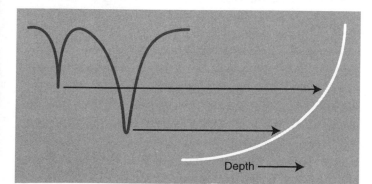

Depth ⟶

Fig. 6.3 The Milne-Eddington explanation of Fraunhofer lines and how they map the photospheric temperature. On the left are two Fraunhofer lines, dips in the background continuous spectrum. On the right is shown a graph of the temperature, increasing inward in the sun. The central intensities of the two lines correspond to the black body radiation at two different depths in the photosphere.

falls to about 4,400 kelvin about 500 km above the level from which most of the photosphere's radiation escapes.

THE GREAT LEAP UPWARD

The next great advance began at the total solar eclipse of 1869. Charles Young, then a professor of astronomy at Dartmouth College, observed a bright emission line (the reverse of an absorption line) in the green part of the coronal spectrum. It corresponded to no known element, and Young thought he had discovered a new element, "coronium." During the following seventy years, twenty-three additional bright lines were observed in the coronal spectrum at eclipses and none of these could be identified with known substances. The puzzle was solved only in 1940.

Until then, everyone thought the corona was cooler than the photosphere—a natural assumption to make since the temperature was known to decrease outward in the photosphere. However, Walter Grotrian, an astronomer at the Potsdam Observatory, began to think otherwise.

His observations showed that although the continuous coronal spectrum resembled the photosphere's spectrum very closely, all the Fraunhofer lines were missing. Grotrian knew that electrons scatter photons of all wavelengths equally well. So to explain why the coronal and photospheric spectra were similar, he suggested that free electrons in the corona scatter light from the photosphere. (This turns out to be completely correct.) But how to explain the absence of Fraunhofer lines?

Grotrian noticed that at the wavelengths where two very deep absorption lines appear in the photospheric spectrum, only a single, very broad and shallow depression appeared in the coronal spectrum. That was the clue he needed. He guessed that he was observing an extremely *hot* corona, in which the rapid motion of scattering electrons smeared out the narrow Fraunhofer lines by the Doppler effect. To account for the broad, deep depression, the corona would have to have a temperature of several hundred thousand kelvin.

A little earlier, Ira Bowen, an astronomer at Cal Tech, had solved the puzzle of "nebulium." Tenuous clouds of gas surrounding hot stars (gaseous nebulae) were observed to emit bright spectral lines that could not be identified with any terrestrial element. Astronomers drew the obvious conclusion that a previously unknown element, "nebulium," must exist. But Bowen showed that under the extreme conditions of a nebula, oxygen and nitrogen atoms would emit spectral lines that were ordinarily absent in laboratory experiments. Was the same sort of thing happening in the tenuous corona?

In the late 1930s Bengt Edlèn, a Swedish spectroscopist, was subjecting various elements to a very hot electrical spark and observing their extremely short wavelength spectra. From these spectra he could work out the energy levels of the different ions he produced. Grotrian read Edlèn's papers, did a short calculation, and uncovered some interesting coincidences.

He used Edlèn's published energy levels to show that a well-known coronal line at 637.4 nm might be emitted by iron atoms that had lost nine of their twenty-six electrons, labeled Fe X. Another line, at 789.2 nm, might be emitted by ten-times ionized iron or Fe XI. The problem was that such spectrum lines were strongly forbidden by the rules of quantum mechanics. Grotrian, a cautious researcher, probably realized his result might only be a numerical coincidence. But he did get in touch with Edlèn anyway and Edlèn was inspired.

To understand how Grotrian made his discovery, and how it led to solving the coronium puzzle, we need to recall how atoms are constructed and how they interact with photons.

ATOMS AND LIGHT

Quantum mechanics is one of the greatest triumphs of twentieth-century physics. While it is an extremely complex subject, we can naively picture one of its consequences as follows: an electron in an atom can only occupy certain distinct orbits around the nucleus. The larger the orbit the larger is the electron's energy. If the electron jumps from a high orbit to a lower orbit, the atom will emit a photon—a particle of light. The photon's energy is precisely the difference of energy of the two orbits. (The picture of orbits, like planetary orbits, is only that—a convenient picture—but the idea of *energy levels* conveyed by the picture holds true.)

Figure 6.4 is an illustration of the electron orbits and the corresponding energy levels of the simplest atom, hydro-

Fig. 6.4 Electron orbits
and corresponding
energy levels in
hydrogen atoms.

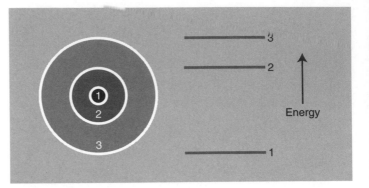

Fig. 6.4 Electron orbits and corresponding energy levels in hydrogen atoms.

gen. If, for example, the electron jumps from level two to level one, it emits a photon with an energy of 10.2 electron volts (ev) or a wavelength of 121.6 nm.

The process works backward, too. If the atom is hit by a photon with exactly 10.2 ev , its electron can be boosted from level one to level two. *This is the origin of spectral lines.* The emission (or absorption) of a photon with a very specific wavelength corresponds to the downward (or upward) jump of the electron between any two levels. The specific wavelengths are atomic fingerprints—they identify an atom as hydrogen or helium or some other element.

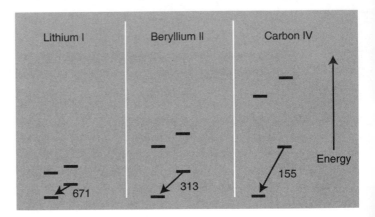

Fig. 6.5 The energy levels in an iso-electronic sequence of neutral lithium, singly-ionized beryllium, and triply-ionized carbon. Notice how the arrangement of levels is similar while the energy scale increases from ion to ion.

The heavier an element is, the more electrons its atoms have. Magnesium atoms have 12, iron atoms 26, gold 79. But in hot plasma where collisions among atoms and free electrons are violent, an atom can lose one or more electrons of its original standard equipment and become an "ion," from the Greek word for "going." The main point about an ion is that its energy levels and its spectrum look nothing like its parent atom's. An ion is almost like another atom.

To see how Edlèn went about identifying the energy levels of ions, and then identifying the mysterious coronium lines, turn to endnote 6.1.

Figure 6.6 shows some of the lowest energy levels that Edlèn found in Fe X, an iron atom that has lost nine of its twenty-six electrons. You should understand that he measured only the strong lines at 9 nm, which are the only ones permitted

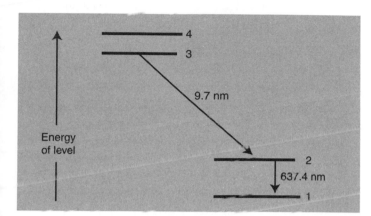

Fig. 6.6 The lowest energy levels in Fe X. The forbidden spectral line at 637.4 nm arises from an electron jump between the two lowest levels.

by the rules of quantum mechanics. But Grotrian noticed that the "forbidden" electron jump between levels one and two would produce the well-known red coronal line at 637.4 nm.

What does it mean to say a line is "forbidden" and yet appears in the coronal spectrum? "Forbidden" is perhaps too strong a term to use in this connection, although it is the preferred one. It is more accurate to say such lines are extremely weak compared to the permitted lines. The reason they are seen in the corona and not in the laboratory is that the corona is extremely tenuous. In the low plasma density of the corona, an electron can occasionally make the forbidden jump from level two to level one before an atomic collision boosts the electron to level three.

Once Edlèn grasped Grotrian's remarkable discovery, he set to work. He systematically determined the energy levels of a large number of ions of abundant elements like iron, nickel, calcium, and silicon, using the regularities of iso-electronic sequences. Then he looked for the kind of energy difference coincidences that Grotrian had found. When he was done he had identified nineteen of the twenty-four known coronal lines. They were all "forbidden" lines. Young's green coronal line at 530.3 nm turned out to arise in thirteen-times ionized iron, Fe XIV. Coronium took its place among other scientific misconceptions such as nebullium, phlogiston, and the luminiferous ether.

The next step was to determine the temperatures at which such highly ionized atoms can exist. Edlèn's spark probably produced temperatures as high as several million kelvin, but he had no reliable way to measure them. Only in the late 1940s, after detailed calculations had been made, was it possible to identify a particular ion (say, Fe X or XIV) with an optimum plasma temperature. Many scientists participated in this enterprise. We'll review some of their ideas in chapter 11.

THE EXTREME ULTRAVIOLET SPECTRUM

Once the high temperature of the corona was established, astronomers were eager to explore the intermediate layers between the photosphere and corona. They would presumably have temperatures between ten thousand and a million Kelvin and their multiply-ionized atoms would radiate in the extreme ultraviolet (EUV), at wavelengths below about 100 nm.

Fortunately for us, the earth's atmosphere blocks all ultraviolet light and protects us from radiation burns. But this benefit to sunbathers is a disadvantage to the astronomer. To explore this regime in the ultraviolet physicists needed rocket technology to carry their instruments above the atmosphere. After World War II American physicists began to load spectrographs aboard captured V2 rockets and to observe the ultraviolet solar spectrum with them. Richard Tousey and Herbert Friedman, scientists at the Naval Research Laboratory, were the first to obtain hints of the ultraviolet solar spectrum.

Further progress had to wait until the late 1960s and early 1970s when the National Aeronautics and Space Administration launched a series of eight Orbiting Solar Observatories. As many as six different instruments, built and operated by independent groups, could be flown on each Observatory. These satellites mapped the broad outlines of the extreme ultraviolet spectrum.

Figure 6.7 is a small portion of the EUV spectrum. The smooth, continuous spectrum of the photosphere trails off below 170 nm and is replaced by a forest of bright emission lines from such ions as carbon II, III, and IV; nitrogen III,

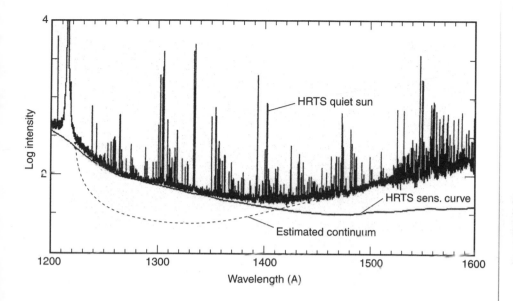

Fig. 6.7 A portion of the solar extreme ultraviolet spectrum. (Courtesy of the Naval Research Laboratory)

IV, and V; and oxygen III, IV, and V. In addition, the strong lines of hydrogen and helium dominate the spectrum below 122 nm. These lines revealed the existence of a new region of the sun's atmosphere, the transition zone, between the chromosphere and the corona, with temperatures ranging between 50,000 and 500,000 kelvin.

The spectrum lines also contain information on the plasma density (see endnote 6.2). Early analyses of the intensities of the EUV lines showed that the transition zone is quite thin—a few thousand kilometers at most—and has plasma densities in the range from 10^{10} to 10^8 protons per cubic centimeter.

The spatial resolution of most instruments aboard these Orbiting Observatories was limited to some tens of thousands of kilometers. As later observatories gained in resolution, they revealed more and more structural detail. Astronomers knew from their ground-based observations that the transition zone was probably not a uniform spherical shell and they clamored for still higher resolution.

NASA answered their pleas with the Skylab Mission (1973–74). An imaging spectrograph, built by a group based at Harvard College Observatory, showed structural details of the chromosphere and transition zone as small as five arcseconds or 3,500 km. Moreover, the Harvard instrument produced simultaneous images in six spectral lines that spanned the temperature range from 30,000 to a million

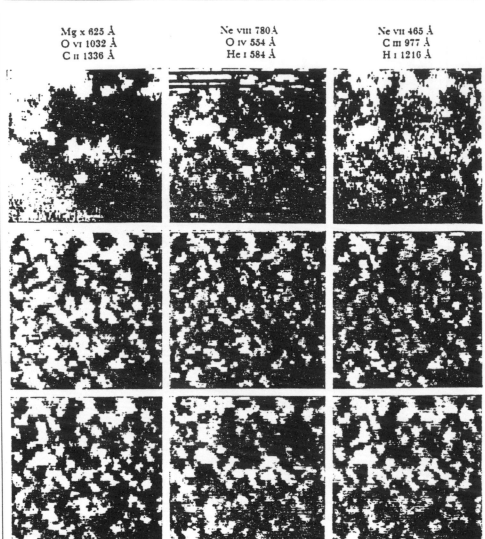

Mg x 625 Å
O vi 1032 Å
C ii 1336 Å

Ne viii 780 Å
O iv 554 Å
He i 584 Å

Ne vii 465 Å
C iii 977 Å
H i 1216 Å

Fig. 6.8 Images of the chromosphere in nine spectral lines, obtained by the Harvard spectrometer aboard Skylab. Each line samples a different plasma temperature. Each blob corresponds to a columnar structure whose temperature increases with height above the photosphere. The structures are similar in all lines except the coronal line of Mg X. (Courtesy of Phil. Trans. R. Soc. London)

kelvin. Another form of imaging spectrograph, built by the Naval Research Laboratories, was a powerful companion to the Harvard instrument. For the first time astronomers could trace the radial variation of temperature to the inner corona in structures of different kinds (Figure 6.8).

Skylab also carried an X-ray telescope, built by the American Science and Engineering Company. Its special filters isolated a narrow wavelength band around 4 nm, that is emitted by plasmas as hot as 3 million kelvin. This telescope revealed the delicate X-ray loops that thread through and connect different active regions.

Skylab produced an avalanche of new science. We shall return to its revelations about the corona in chapter 11.

Putting It All Together: Empirical Models

We have seen how the emission and absorption spectra of the solar atmosphere contains information on the atmosphere's temperature and density. The goal facing astronomers in the late 1970s was to extract this information and incorporate it in a numerical "model atmosphere" that would represent the sun. Such a model, when combined with the appropriate atomic physics, would accurately predict all the observed lines and continua, in the X-ray, ultraviolet, visible, infrared, and radio spectrum.

Skylab had shown that even the "quiet" sun was not perfectly uniform or homogeneous. The borders of supergranules were brighter in some chromospheric lines than their centers, for example. Obviously it would be necessary to build separate models for different parts of the quiet atmosphere.

Constructing even a single model is a huge task, because it involves calculating the profiles of hundreds of spectrum lines and dozens of continua, from the extreme ultraviolet to the far infrared. It requires knowing the sources of opacity at every wavelength. Such a job was only feasible when fast computers came along.

Early attempts were made by Owen Gingerich and his colleagues at Harvard College Observatory in 1971 and by Dutch astronomers at the Utrecht Observatory. Eugene Avrett and his associates Jorge Vernazza and Robert Loeser carried out the most complete calculations at Harvard University, after the Skylab Mission.

Avrett doesn't fit the contemporary image of an academic. No jeans, sandals, or ponytail for him. He is serious in demeanor, precise in speech, and conservative in dress. He has gained a reputation for meticulous work and for the courage to tackle difficult tasks. He began his tremendous labors in the early 1970s and has spent the following twenty-five years incorporating new data and new physics into his models.

Avrett and friends began by searching for a single temperature model that would predict the ultraviolet, infrared, and millimeter continuum radiation from the center of the quiet solar disk. They assumed that the atmosphere is in hydrostatic equilibrium, a reasonable assumption at least for the lower layers. They found such a model in 1976.

Then they modified this basic model to produce a range of temperature models. By a process of trial and error they found models that would successfully predict the observed wavelength profiles of visible and ultraviolet spectrum lines, in several types of atmospheric structures. To do this they had to determine how the opacity of the solar plasma varies, over a huge range of wavelengths. And they had to compute thousands of individual line profiles as a check. They completed this task in 1981, seven years after the end of the Skylab Mission. Their final temperature model for the "average sun" is shown in Figure 6.9.

Fig. 6.9 The temperature profile of the "quiet" sun, determined from a wealth of spectra. (Courtesy of E. Avrett)

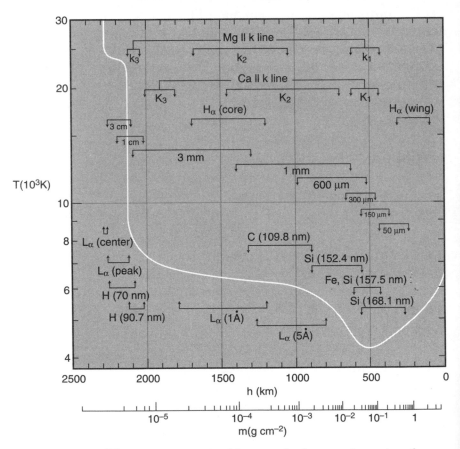

The temperature in this smoothed atmosphere rises from a minimum of 4,700 kelvin, slowly at first in the chromosphere and then very steeply in the transition zone. To predict the shape of the strongest ultraviolet line of hydrogen (the Lyman alpha line , Lα, at 121.6 nm) they had to introduce a somewhat artificial temperature plateau at 25,000 kelvin.

Notice the various labels in Figure 6.9 that indicate where in the atmosphere certain lines and continuous radiation are emitted. Most of the information on the temperature minimum region comes from the far infrared spectrum of the center of the solar disk.

Once their empirical model reproduced the observed spectra successfully they could determine the rate at which energy is lost at each height in the low atmosphere. This is the amount of nonradiative energy that must be supplied by some unspecified means to prevent the atmosphere from cooling off. About 5 kilowatts of nonradiative energy must enter each square meter at the base of the atmosphere to balance these losses. Compare that to the 60,000 kilowatts of visible light that escapes the photosphere!

In 1989, Grant Athay and Lawrence Anderson, two scientists at the High Altitude Observatory in Boulder, Colorado, confirmed the work of Avrett and associates using a different approach (see endnote 6.3).

But how is this energy supplied? What actually heats the atmosphere? These are old questions that are still not fully resolved. One idea has been around for a long time, however.

HEATING BY SOUND WAVES

After astronomers recognized in 1940 that the coronal temperature is as high as 2 million kelvin, they began to think of possible sources of the nonradiative energy that heats it. The first and most influential suggestion was made independently by Ludwig Biermann in 1946 and Martin Schwarzschild in 1948. They proposed that sound waves are generated in the turbulent convection zone and propagate into the upper atmosphere. As they encounter lower and lower gas density the waves steepen into shock waves, as ocean waves steepen when they ride up a beach. Eventually the shocks will dump their acoustic energy into the gas and heat it.

Observers immediately set out to find such sound waves in the photosphere, but without a clear idea of their periods. Many theorists began to ask how waves might propagate and where in the atmosphere they might dissipate as shocks, but without a guide to the amount of power they must carry.

The idea of acoustic heating held the attention of astronomers for decades. But before a quantitative theory could be advanced, several essential questions had to be answered. How much acoustic energy does solar convection generate? How much can reach the atmosphere? Which sonic frequencies are involved and how should observers search for them? Where will the acoustic energy dissipate and can it supply energy where it is needed? Will enough reach the corona to maintain it at 2 million kelvin?

Sir James Lighthill, a British applied mathematician, offered an answer to the question of noise generation. Lighthill was interested in all things involving fluid dynamics and was one of the first scientists to examine how fish swim and to model traffic flow. An obituary described him as an "inveterate circumnavigator." In his youth he was the first person to swim around Sark, one of the Channel Islands off the coast of France. At age seventy-four, he tried to repeat this achievement, but unfortunately drowned in the attempt.

Lighthill was concerned with the amount of sonic noise produced by jet engines. Not only was the noise draining energy from the engines but it obviously annoyed listeners. In 1952 he published a theory of noise generation by turbulence in a homogeneous atmosphere and experiments carried out later showed his theory gave good predictions.

Astronomers wholeheartedly embraced Lighthill's theory as the first to make verifiable predictions. They extrapolated his work to the solar convection zone, although it is far more turbulent than any jet engine. In 1967, Robert Stein extended Lighthill's theory to apply to an atmosphere with strong height variations of density, a much more realistic model for the convection zone. With many uncertain assumptions he was able to predict that solar convection would produce sound waves primarily with periods between twenty and sixty seconds. He estimated an energy flux of 100 kilowatts per square meter would reach the photosphere, but cautioned that he could easily be off by a factor of ten either way.

With some guidance on the periods of sound waves the convection zone produces, theorists could begin to refine their models of wave propagation and dissipation. Peter Ulmschneider at the University of Heidelberg and his former student Wolfgang Kalkoven have been among the most prolific. Beginning in 1970 and continuing to the present day they've constructed more and more elaborate calculations of the propagation of sound waves and the formation of shocks. Others have participated as well and so these aspects of the theory have advanced greatly.

But observations will generally prevail over theory. A fatal blow was dealt to the whole idea of acoustic heating by R. Grant Athay and Oran Richard White, from the High Altitude Observatory in Colorado. In 1978 they obtained spectra at ultraviolet wavelengths near 180 nm with an instrument aboard Orbiting Solar Observatory Seven. From the breadth of the line they estimated the maximum amount of acoustic energy flowing through the middle chromosphere, at a height of about 2,000 km. They concluded that too little sonic power was available to heat the upper chromosphere and the corona. What about the lower levels of the atmosphere, however? (See endnote 6.4.)

Then, after instruments aboard the Skylab Mission showed a close correspondence between bright emission in ultraviolet lines and magnetic field strength, the majority of astronomers were convinced that acoustic heating is unimportant, at least for the chromosphere and corona, and turned to magnetic sources of heating. So it is time to examine the sun's fields.

Chapter 7

THE MAGNETIC
ATMOSPHERE

 **I**f the sun's atmosphere were built like an onion, with one smooth layer over another, our journey would soon be over. If all we could see was the rise of temperature to the outer layer, the corona, our tale would soon be told. But take a look at Figure 7.1, a view of the corona from the SOHO satellite in extreme ultraviolet light. The

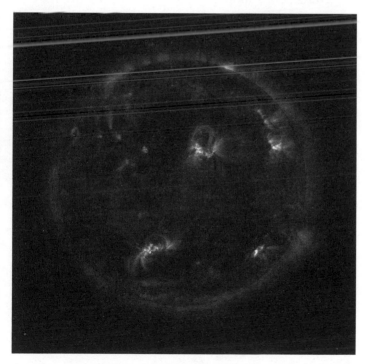

Fig. 7.1 An image of the sun in extreme ultraviolet light that is emitted by coronal plasma at a million kelvin. Coronal loops, shaped by the magnetic field, are visible above active regions. Also see color supplement. (Courtesy of SOHO/EIT consortium)

familiar smooth face of the sun is covered with loops and arcades and bald spots of all sizes. Around its edges we see wispy traces of the solar streamers, which show up in all their grandeur during a total solar eclipse (Figure 7.2).

All these complicated structures are shaped by the atmosphere's pervasive magnetic fields. In the SOHO image, hot plasma outlines the lines of force of the field, which

Fig. 7.2 The corona
photographed at the
total solar eclipse of
July 11, 1991. The
streamers are shaped
by plasma flows and
by coronal magnetic
fields. (Courtesy of the
High Altitude
Observatory)

would be invisible otherwise.

At first glance there doesn't seem to be any order to the arrangement. In the lower left quadrant of the SOHO picture we see an intensely bright set of coronal loops that are anchored in a cluster of sunspots. On the far right we see another "activity complex" in profile, with tremendous loops extending far beyond the edge of the sun. At the north pole (at the top) lies a dark blank region, nearly devoid of structure—a coronal "hole." And a messy tangle of loops and arcades is spattered, seemingly at random, over the rest of the solar disk.

If we looked a month later or a year later we'd find that most of the details would be different. And indeed if we watched for a decade we'd see really big changes. At one point the sun would grow almost completely bald with only a fringe of streamers around its equator.

But all this confusion is only apparent. The sun's magnetism is actually highly ordered, both in space and in time. Perhaps the corona is not the best place to find the order, however. The place to start is in the photosphere, where the coronal magnetic fields are rooted, and where observations of magnetic fields are most easily made. So we will shift our perspective momentarily to the photosphere, where we will track the changes of magnetic field over the surface and during the solar cycle. We'll recall a little history along the way.

THE SOLAR CYCLE

Our story begins in 1826 with Samuel Heinrich Schwabe in the German town of Dessau. Schwabe was making a good living as an apothecary but his real love was astronomy. Just as a hobby, he began to record the number of sunspots each day. Within a year he noticed that the number of spots on the face of the sun was growing rapidly. It reached a peak in 1830 and then, surprisingly, started to decline. Schwabe was intrigued, and being a methodical man, continued his counts for the next thirteen years. He was then able to announce that the number of spots varied in a cycle of ten years. He was a bit off—later counts established eleven years as a good average, although individual cycles vary from eight to fourteen years.

Richard Carrington, the son of the rich brewer whom we met in connection with the rotation of the photosphere, made the next important discoveries. After the number of spots reaches a minimum, new spots appear on both sides of the equator at latitudes around 30 degrees. As the cycle advances, new spots appeared at lower and lower latitudes in two belts between latitudes 10 and 30 degrees. The last spots of a cycle emerge at latitudes of less than 10 degrees.

The British astronomer E. Walter Maunder illustrated Carrington's findings in the now famous "butterfly diagram." Figure 7.3, top shows these patterns for cycles beginning in 1880. Follow any cycle and you will see the be-

Fig. 7.3/top: The "butterfly" diagram of the latitude distribution of sunspots during the past 11 cycles.

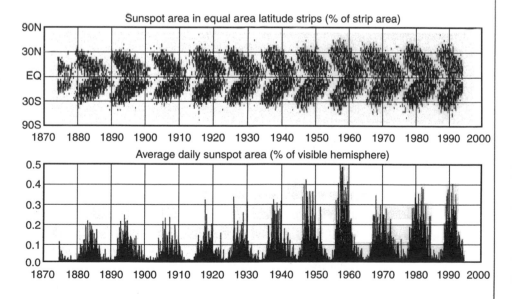

Sunspot area in equal area latitude strips (% of strip area)

Average daily sunspot area (% of visible hemisphere)

havior Carrington noted: a slow start at high northern and southern latitudes, increasing numbers of spots at lower and lower latitudes, and a final fizzle almost at the equator.

The lower panel, 7.3, bottom, displays the area covered by sunspots during a cycle. Even in the best of times the spots

Fig. 7.3/bottom: The area (in percent of the solar disk) covered by sunspots during the cycle. (Courtesy of D. Hathaway)

cover no more than half a percent of the solar disk. Their rise to maximum takes an average of 3.9 years, the decline to minimum 7.0 years.

This diagram has been analyzed to exhaustion, as people have tried to correlate the price of wheat, the length of women's skirts, the frequency of Republican Congresses, the quality of French wines, and the weather, to the fluctuations in sunspots. An eighty-year Gleisberg supercycle that modulates the maxima of the cycles does seem to be present, but the length of the "eleven-year" cycles seems to vary randomly. Also, the northern and southern hemispheres can behave quite differently in the same cycle; for example, look at the butterfly centered at 1970.

Perhaps the most essential feature of sunspots, their magnetic fields, was reserved for George Ellery Hale to discover. A towering figure in twentieth-century astronomy, he was a master of establishing new institutions, such as the Yerkes Observatory, the Mt. Wilson Observatory, the Palomar Observatory, and the *Astrophysical Journal*. He also played a key role in founding the International Astronomical Union, and the California Institute of Technology. Despite all his entrepreneurial activity, he found time to excel as a scientist, and was interested no less in solar physics than in the structure of the universe.

Around 1907, the powerful new Snow telescope at the Mt. Wilson Observatory revealed that some spectrum lines in sunspots are doublets: two lines appear where only one is expected. Soon afterward, images of sunspots were made in the strong spectrum line of hydrogen at 656.3 nm, the "H alpha" line. These images showed remarkable vortices surrounding sunspots, like pinwheels. They led Hale to think that spots might rotate and thus accelerate free electrons to produce a magnetic field. He was also familiar with the research of Pieter Zeeman, a Dutch physicist who had discovered that spectrum lines split into doublets when their radiating atoms reside in a strong magnetic field. Zeeman had also found that the light of each line of the doublet is polarized in the opposite sense to its partner, like left- and right-handed screws. In a stroke of intuition, Hale tested the sunspot doublets for polarization and discovered that indeed the spots contained magnetic fields as strong as 3,000 gauss. (For comparison, the earth's magnetic field is less than half a gauss.) Later research has shown that only some spots look like vortices and that none really rotates. So, following a false chain of reasoning, Hale nevertheless made a monumental discovery.

Hale and his colleagues at Mt. Wilson went on to discover several basic regularities among the magnetic polarities of sunspots during the sunspot cycle. They show up nicely in recent maps of photospheric magnetic fields, Figures 7.4 and 7.5, which are typical of the minimum and maximum of the cycle, respectively. In the figures, positive fields (pointing out of the sun) are shown in white, negative fields in black.

According to Hale, spots usually emerge from below the photosphere in pairs, with opposite magnetic polarities. Each pair forms a magnetic "bipole," whose axis (the line joining the spots) is aligned approximately east-west on the solar disk. Notice the pairs of white and black areas in Figure 7.5 (a good example is at 10 o'clock), which are the active regions surrounding clusters of opposite polarity sunspots.

Recent research reveals that new sunspots tend to emerge near older sunspots, and cluster in an "activity complex" that may last for many months. Many complexes are visible in Figure 7.5.

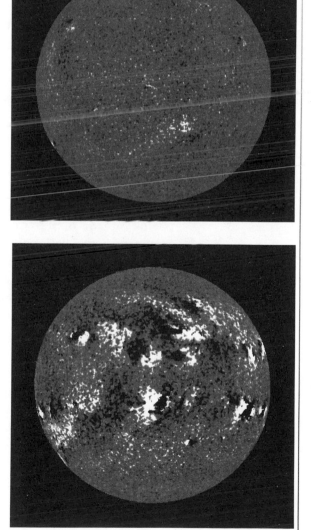

Fig. 7.4 A magnetogram showing the magnetic fields of opposite polarities, all over the sun near solar minimum.

Fig. 7.5 Another magnetogram, corresponding to a period of high magnetic activity, March 2, 2000. (Courtesy of National Solar Observatory)

Next, Hale tells us that in the northern hemisphere, all the westernmost spots, or "leaders," have the *same* magnetic polarity, which is opposite to the polarity of leader spots in the southern hemisphere. Figure 7.5 shows this nicely. North is at the top, west on the right, and each western half of an active region is positive in the north and negative in the south.

But that is not the whole story. In the next cycle, all the bipoles will reverse polarities. The leaders in the north will then have negative polarity and their companions ("followers") will have positive polarity. The full cycle is therefore twenty-two years, not eleven.

Hale established the study of solar magnetism as a major program at the Mt. Wilson Observatory, but until the early 1950s the work was carried out with Hale's basic method, photographic spectroscopy. The method was slow and tedious and limited to the strong fields in and around sunspots.

By the early 1950s, technology had advanced to the point that a more sensitive instrument was feasible. Harold and Horace Babcock, a father and son team at Mt. Wilson Observatory, decided to build one. Harold had retired in 1948 after a productive career studying the sun. He encouraged his son Horace, a new member of the staff, to invent a new instrument, the solar magnetograph, to measure the weak fields outside of activity complexes. Horace was well equipped with a background in optics, electronics, and astrophysics. After several years of work, he completed the first model. It measured the polarization of a sensitive photospheric line and displayed the resulting magnetic field strength and polarity on a cathode-ray tube.

With their magnetograph, the Babcocks discovered that activity complexes extend much further over the solar surface than anyone had realized. Moreover, both weak and strong photospheric fields coincide with bright emission in chromospheric spectrum lines, like the calcium K line, at 393.3 nm. In fact, an image made in the light of the K line (a spectroheliogram) is a fairly good map of the strength of photospheric fields.

The Babcocks were the first to measure the weak field near the poles of the sun. The poles have opposite magnetic polarities and, before sunspot maximum, each pole has the same polarity as the leaders in the same hemisphere. (You can see this effect in Figure 7.4, although faintly.)

The Babcocks found that the polar polarities also reverse, within a few months of sunspot maximum. Thomas G. Cowling, the English theorist we've met before, pointed out that this could not happen in so short a time if the sun's field resembled the earth's, with field lines passing from pole to pole through the interior. The large-scale field must be confined therefore to relatively shallow layers.

In Figure 7.5, the strong-field map, you will notice diagonal stripes of alternating polarity at high northern latitudes. The Babcocks found these in their magnetic maps and called them "bipolar magnetic regions." They trail out of mature active regions and evidently represent a magnetic field that has spread from those regions. (We need to keep in mind that solar magnetic fields are basically *loops or arches*, that connect patches of opposite polarity on the photosphere. A magnetic map like Figure 7.4 only shows the *roots* of these loops, like the bristles on a brush. The full three-dimensional loops only become obvious in an X-ray or extreme ultraviolet picture, like Figure 7.1.)

It remained for Robert Leighton, the discoverer of supergranules, to explain how the Babcocks' bipolar magnetic regions form during the decay of an active region. Just below the photosphere, the horizontal flow in a supergranule is strong enough to displace the roots of a magnetic loop. As a series of supergranule cells wells up at the borders of a mature active region, they peel off magnetic loops and pass them to supergranules further away. These cells then shuffle the loops further and further into the surrounding quiet photosphere. The process resembles the diffusion of ink in water; a single drop if left alone will slowly spread throughout the water because of random motions of water molecules.

A single root of a magnetic loop traces a "random walk" over the photosphere. Like a drunk walking home, the root takes a random step (half the diameter of a supergranule, say, 15,000 km) in any horizontal direction, once during the lifetime (about a day) of the supergranules. At the same time supergranules disperse the active region field, differential rotation (the equator rotates faster than the poles) drags the field footpoints into the diagonal bipolar patterns we see in Figure 7.5.

The dispersed fields eventually cover the whole sun. Leighton and his student George Simon demonstrated that the supergranules concentrate the fields they have stolen from active regions into a lacy "network" that outlines their borders (Figure 7.6). Moreover, this network coincides with

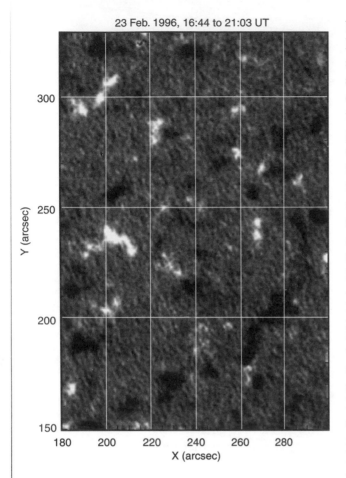

23 Feb. 1996, 16:44 to 21:03 UT

Y (arcsec)

300

250

200

150

180   200   220   240   260   280

X (arcsec)

Fig. 7.6 Supergranular cells concentrate magnetic flux of both positive (white) and negative (black) polarity at their borders in a so-called "coarse network." The borders are also the sites of bright chromospheric emission and strong downdrafts. (Courtesy of SOHO/MDI consortium)

bright chromospheric line emission (indicating a hot, dense chromosphere) and also coincides with downdrafts at the borders of the supergranules. These three clues fit a consistent picture of convective motions dragging the roots of magnetic fields to their borders, where processes as yet unclear heat the plasma. More on this later!

Neil Sheeley (another Leighton student) and his colleagues at the Naval Research Labs have demonstrated with numerical simulations how Leighton's diffusion process leads to the cyclic reversal of the sun's polar fields. Two essential ingredients are necessary: meridional flow (a steady 10 m/s flow of photospheric plasma to the nearest pole) and the tilt of the axis of a pair of sunspots.

When a pair of spots emerges, the follower spot is usually closer to the pole, for reasons that are just becoming clear. As the fields disperse, meridional flow sweeps both polarities toward the pole, but because of its original advantage the follower polarity (which is opposite to that of the pole) reaches the pole before the leader's polarity does. When the follower's field arrives at the pole it cancels the weaker polar field and replaces it with its own polarity.

It's time to summarize. The solar cycle is a magnetic cycle of eleven years in which the average number of sunspots visible over the surface varies from a minimum near zero to a maximum of about a hundred. Individual spots disappear in a day or a week, but are replaced to maintain the average number. At the beginning of the cycle, spots appear in both hemispheres at about 30 degrees latitude and as the cycle progresses, they appear in greater numbers at lower latitudes. They obey Hale's laws, with polarity reversal of leader and follower spots and of the poles, in successive cycles. According to Leighton's field transport model, active region fields are dispersed by supergranule motions and dragged out into diagonal bipolar magnetic regions by differential rotation and a slow poleward flow.

This neat scenario has been established observationally beyond any quibble, but every part of it cries out for an explanation. How does the sun generate its magnetic field and why do sunspots choose to appear according to the butterfly diagram (Figure 7.3)? How do we account for Hale's laws? And most important, what happens to all the field that emerges during a cycle? (Measurements show that the total amount of magnetic "flux" [lines of force] varies by a factor of *fifteen* during the cycle.) We'll tackle these problems later in chapter 12.

The solar cycle concerns much more than sunspots or bipolar regions. Everything we see in the atmosphere of the sun varies in step with the cycle. For example, the size and location of active regions, the latitudes of coronal streamers, and the frequency of flares all vary in step with the eleven-year cycle. The reason, obviously, is that all these phenomena are driven by the changes in the atmosphere's magnetic field.

Well then, does *everything* on the sun change with the sunspot cycle? Is nothing constant? Until a few years ago astronomers would have said at least one thing is rock-steady: the amount of energy the sun radiates into space. They even coined a label for the fraction of this energy the earth receives: the Solar Constant.

THE SOLAR CONSTANT?

Charles Greeley Abbot, the second director of the Smithsonian Institution in Washington, devoted a large part of his career to measuring the amount of energy the earth receives from the sun.

From 1906 to about 1920, he conducted a continuous series of observations using a bolometer, an electrical instrument that Samuel P. Langley had invented. In order to minimize the absorption of sunlight in the earth's atmosphere he established a number of small observatories on mountaintops in the U.S. and in Peru. But even at an altitude of 14,000 feet atop Mt. Whitney, some absorption remained, and worse, varied with the weather. Despite his best efforts to correct for this systematic effect, his results were uncertain by at least a percent.

Although Abbot convinced himself that the solar constant actually varies by something like a percent, the consensus among astronomers is that his corrections for absorption corrupted his results. They concluded that only observations from space, with a very sensitive instrument, could reveal a real variation if indeed any exists.

That opportunity arose in 1980. In that year Richard Willson, a physicist at the Jet Propulsion Laboratory in Pasadena, installed his Active Cavity Radiometer Irradiance Monitor (ACRIM, for short) aboard NASA's Solar Maximum Mission satellite. This device was capable of detecting changes, as small as a few parts in a million, in the amount of solar energy reaching the earth. Since then several similar devices have been flown, the latest (VIRGO) aboard the SOHO satellite. Two whole cycles have been monitored.

Figure 7.7 displays the combined results from all the instruments. The solar constant actually varies by 0.3 percent during the eleven-year sunspot cycle, with a brighter sun at sunspot maximum. Abbot was right after all!

The appearance of a single large sunspot can cause a 0.2% dip in the curve. Evidently a dark spot can temporarily reduce the amount of energy the sun radiates. That energy must be stored as heat in the interior because after a few days it reappears in the ACRIM record.

How can we explain the eleven-year variation of the solar constant? Theorists tell us that the sun generates energy in its core at a rate that changes only over millions of years, so the eleven-year cycle must be produced somehow near the surface.

According to current ideas, most of the energy blocked by sunspots is radiated within a few days from the bright chromosphere above active regions and above the magnetic network. However, the chromospheric emission and sunspot absorption are not precisely equal. A fraction of the blocked energy is apparently stored as heat in the convection zone for years and released only over the eleven-year

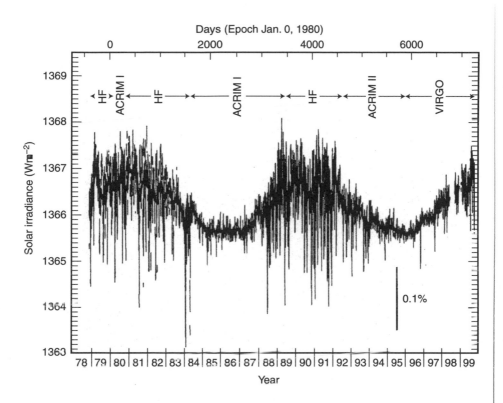

Days (Epoch Jan. 0, 1980)

Fig. 7.7 Two solar cycles of observations of the so-called solar constant. These are data combined from several instruments including ACRIM I and II, and those presently aboard SOHO. (After Frohlich and Lean. Courtesy of SOHO/ VIRGO consortium.)

cycle. That could account for the tiny variation of the solar constant.

Most of the energy the earth receives lies in the visible part of the solar spectrum, and this part varies by only 0.3% during the eleven-year cycle. But the short wavelengths in the solar spectrum, which contribute only a few percent of the total energy, fluctuate wildly during the cycle. The flux of soft X-rays, for example, can vary by a factor of ten.

THE HUNTING OF THE FLUX TUBE

So far we've discussed only the "big picture"—the large-scale, long-term ordering of the sun's magnetic field. But there are more surprises to come.

The late 1960s and early 1970s introduced a revolution in our views of the solar magnetic field. Formerly all the emphasis was on the big concentrations of field-sunspots and active regions. But as technology improved and more sensitive equipment became available, astronomers realized that the solar magnetic field is organized into a *hierarchy* of sizes and strengths, and that the lower limit of sizes could lie below the resolving power of the best telescopes.

In order to appreciate these results we need to distinguish between field strength and flux. The *strength* of the field, measured in gauss, determines the force the field exerts on a moving charged particle, like an electron. (Or if you prefer, on a bar magnet.) The *flux* is a measure of how many magnetic field lines pass through an area. It equals the product of the field strength times the area and is measured in maxwells.

As an example, suppose a sunspot has a field strength of 3,000 gauss, and an area of a hundred million square kilometers or 10^{18} square centimeters. Then it contains a flux of 3×10^{21} maxwells.

The bristles on a hairbrush give one a useful analogy to distinguish field strength from flux. Think of the bristles as the roots of magnetic fields in the photosphere. Then the spacing of the bristles corresponds to the field strength and the total number of bristles in the brush to the flux. The two quantities, strength and flux, are independent. A loop of magnetic field with a certain flux may have either a strong or weak field strength depending on whether its lines of force are packed closely or loosely.

With that as background let's proceed with our story.

For many years astronomers have made images of the solar chromosphere in the light of the strong chromospheric line of ionized calcium, the "K" line at 393.3 nm. Figure 4.2 is an example. It shows bright emission in active regions as well as in the network at the borders of supergranule cells. Under good observing conditions the bright areas in such images resolve into small blobs, around a thousand kilometers in diameter. After the Babcocks showed that bright calcium emission coincides with strong magnetic fields, it seemed obvious that the magnetic roots in the photosphere should also resolve into such small blobs. But how to prove this? Even today, magnetographs rarely resolve such small magnetic patches.

In 1968, Neil Sheeley and his student Gary Chapman photographed the photosphere near active regions and showed that granules are ringed with a lacy pattern of bright emission, a much more delicate network than Leighton's "coarse" supergranule network. This was a clue that magnetic fields in the photosphere do cluster into tiny clumps. But Sheeley and Chapman couldn't measure the fields in their photospheric network.

In the same year, Jacques Beckers and Egon Schröter obtained highly resolved polarized spectra of an active region, which revealed that the photospheric magnetic field is concentrated in otherwise invisible "magnetic knots," a thousand kilometers across, that cover only a few percent of the area of the active region. They estimated their field strengths at a thousand gauss.

Meanwhile, Sara Martin and Karen Harvey, two indefatigable researchers at the Big Bear Solar Observatory, studied the pygmies among active regions. Unlike sunspots, these tiny bipoles pop up nearly uniformly all over the sun and live for only a few hours before dissolving. They contain about a thousand times less magnetic flux than a large active region. They are appropriately named "ephemeral active regions" or EARs. Like sunspots, they obey Hale's laws of polarity and their numbers follow the eleven-year cycle, reaching a minimum about a year before the spots. At maximum about a thousand of these little regions cover the sun.

But even these midgets are not the smallest bipoles.

In 1975, Jack Harvey (Karen's husband) and Bill Livingston, astronomers at Kitt Peak National Observatory, discovered the "intranetwork fields." These magnetic bipoles pop up in the interiors of supergranule cells, and contain about a hundred times less flux than the EARs. According to Sara Martin the two polarities separate and are swept into the network in less than a day. These microscopic creatures emerge oriented at random, and disobey Dr. Hale's laws of polarity. The total amount of field they represent is uncertain, as is their cyclic behavior. They may very well ignore the sunspot cycle entirely and just continue to break out all over the face of the sun, like acne. According to Harvey and Livingston they are so closely packed that, when the two polarities are colored white and black, "the pattern looks like pepper and salt."

By the early 1970s astronomers were beginning to wonder whether there was any lower limit to the sizes of magnetic elements. Like particle physicists looking for the quark, they asked whether a "fundamental" magnetic flux element, the so-called "flux tube," exists. In 1973, Jan Olaf Stenflo, a Swedish solar astronomer, found evidence that they really do exist and have kilogauss field strengths (see endnote 7.1).

Stenflo didn't actually resolve the individual flux elements, but he could estimate their sizes. Richard Dunn took the next step at Sacramento Peak Observatory. In 1973, he photographed granules in white light with higher resolution than anyone had before. His pictures revealed bright threads (the "filigree"), residing in the narrow lanes *between* granules. They could not be wider than the lanes themselves, perhaps 200 km. They suggested a link with Stenflo's unresolved magnetic elements. Finally, Richard Muller of the Pic du Midi Observatory resolved Dunn's filigree into tiny dots no larger than 150 km, that survive for about 18 minutes. The flux tube was beginning to take shape.

Cornelis Zwaan, the late dean of Dutch solar astronomers, has emphasized that all the magnetic features we have described, from sunspots to flux tubes, are related. They are all examples of bundles of magnetic field, differing only in the amount of flux they carry, that emerge from below the photosphere and migrate over the surface. Indeed, they all may be composed of elementary flux tubes.

How to Explain the Strong Fields?

By 1976 it was becoming clear to theorists that they had a serious problem with these observations. They had no problem in picturing how convective cells spreading just below the photosphere, could sweep magnetic flux into their borders. And it was obvious that the flux would be concentrated in the downdrafts at these borders, to form the coarse network that Leighton and Simon had discovered.

The big problem was how to explain why nearly all the photospheric flux is concentrated in tiny tubes that occupy a few percent of the surface area and have *kilogauss* field strengths. The best a theorist could extract from the downdraft process was about 400 gauss, the strength at which the magnetic pressure inside the flux tube would equal the gas pressure outside.

Back in 1941, Ludwig Biermann had proposed a different *thermal* scenario to explain the strong fields in sunspots (see endnote 7.2). Eugene Parker, an astrophysicist at the University of Chicago, seized on Biermann's idea in 1955 and contributed several new ideas that made it a quantitative theory of sunspots. (Parker has probably contributed more ideas to solar physics than any other theorist since Biermann. We'll run across him repeatedly in this and later chapters.)

In 1978 Parker added a critical ingredient to Biermann's

thermal enhancement of field strength and applied it to flux tubes. He realized that when heat is cut off from the tube, its internal plasma will not simply cool in place, but will *sink*. In fact a *steady downdraft* will be established inside the tube, independent of the downdrafts which surround the tube. The combination of cooling and downflow can concentrate the field dramatically (endnote 7.3).

It's a neat scheme. But does it agree, in detail, with observations? To find out, astronomers have mounted a tremendous effort in recent years to explore the interiors of flux tubes. They are hampered by the fact that the predicted diameters of flux tubes are at or *below* the limit of resolution of even the largest and best solar telescopes anywhere on earth. But as we shall see, flux tubes are not just curiosities; they seem to be essential factors in heating the upper atmosphere. So there are good reasons to investigate their properties and their interactions with surrounding granules. Ingenious schemes have been devised for this purpose.

Many scientists are engaged in the pursuit of the elusive flux tube, but we can mention the work of only a few. We will focus on three very active groups.

INSIDE THE FLUX TUBE

Jan Stenflo, the Swedish scientist we mentioned earlier, has trained several generations of students in his techniques of spectroscopic polarimetry. They approach their task with a definite philosophy. They reason that since there seems little chance of resolving individual flux tubes with present equipment, the best one can do is observe small areas in the photosphere spectroscopically and to interpret the results in terms of clusters of unresolved flux tubes (endnote 7.4).

Figure 7.8 shows a 1993 model of a vertical cylindrical flux tube that seems to explain most of their observations. The tube sits in a downdraft of 4 km/s in the neighboring field-free photosphere. It is squeezed tightly by the pressure of the surrounding plasma. Inside the tube most of the plasma has been evacuated, and since the internal plasma pressure is low, the tube's magnetic pressure must resist the external pressure. Therefore the flux has bunched up to produce a strong magnetic field in a tight constriction. All this agrees with Biermann's and Parker's ideas on the origin of the strong fields.

At the surface of the photosphere ($z = 0$), the diameter of

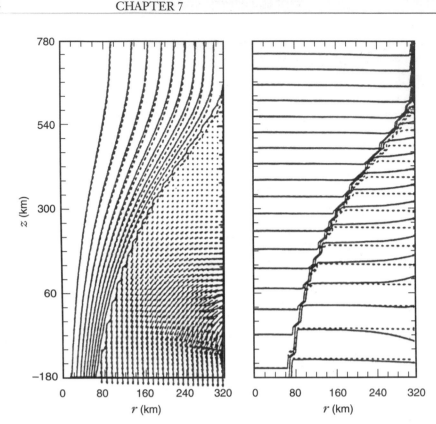

Fig. 7.8 A flux tube model based on observations of a flux tube at the center and edge of the solar disk. The left panel shows half of a cross-section of a cylindrical tube, with magnetic field lines and flow directions drawn in. The right panel shows how the pressure isobars vary in height and width. (Ast. Ap. 268, 746, '93. Courtesy of S. Solanki.)

the tube is 200 km, the uniform field strength is 1,600 gauss, and the tube is cooler than its surroundings. At greater heights, the tube temperature is higher than the surroundings. This height-variation of temperature accounts for the observation that tubes near the limb of the sun are brighter than the surrounding photosphere and are darker at the disk center. Notice how the tube's field lines flare out above the surface. This effect is in response to the lower gas pressure outside the tube at greater heights.

Some of the earlier work of Stenflo's group suggested that a "weak" field (of only 400–800 gauss) of *opposite* polarity to the strong field (1,500–2,000 gauss) is also present in the photosphere. This opposite polarity might represent the flux of the strong field bending over in a loop to return to the photosphere. So we have a preliminary picture of how the magnetic tubes might look in three dimensions.

During the past decade, observers of the tiny flux tubes have been guided by the work of a small number of German theorists. Following the pioneering studies of the Dutch theorist Henk Spruit, numerical hydromagnetic simulations have been made in one, two, or three dimensions, either in static or time-dependent flux tubes or sheets. The aim of the Germans' work has been to describe the full physical state of the tubes, taking into account the forces, motions, and energy balance. With such models they can then calculate synthetic polarized spectral lines for comparison with observations.

Figure 7.9 shows a 1991 model by Michael Knoelker and associates. The tube is modeled as a two-dimensional sheet of magnetic flux in a field-free photosphere. It lies in a downflow of 2 km/s, which helps to constrict the tube. Right at the boundary of the tube the flow speeds up to 6 km/s. The plasma in the tube is highly evacuated and near the surface of the photosphere (z = 0) it is about 200 kelvin

Fig. 7.9 A two-dimensional flux tube model. The panels show the density, field strength, temperature, and plasma velocity in the vicinity of the tube. Only the right half of the tube is shown. (Courtesy of M. Knoelker)

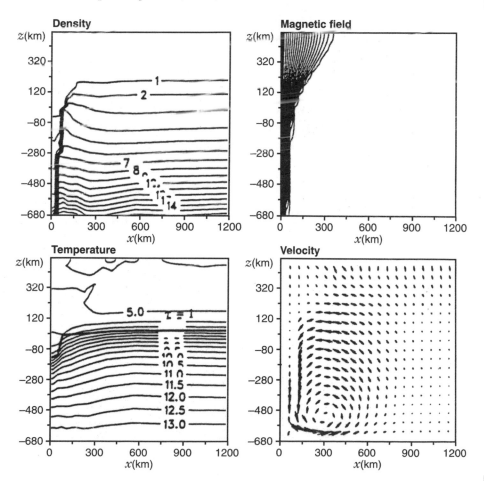

cooler than its surroundings. At a height of 320 km, however, the tube is 200 kelvin hotter. You can easily see how these MHD models have influenced the interpretations of

Stenflo's students and how their observations have guided the theorists.

This time-dependent simulation revealed an unexpected *five-minute oscillation*, in both the tube and in the surrounding photosphere, at least in the initial formation phase. This is a tantalizing indication of the "tube waves" that we will hear more about in connection with the heating of the corona (chapter 11).

LIFE AND DEATH ON THE MAGNETIC NETWORK

We have postponed until now the vexing question of what happens to all the magnetic flux that boils up from below the photosphere. As we have seen, a lot of the flux in active regions apparently survives to migrate for many months over the photosphere. But a large fraction (perhaps most) of the regions' flux, including that of the sunspots, simply disappears in place. Exactly how this happens is still controversial. We'll return to this question in chapter 13, but now we'll focus on what happens on the quiet sun, in the magnetic network. As we'll see, the tiny flux tubes are again in the spotlight.

In 1984 Sara Martin observed how small patches of magnetic flux migrate over the photosphere, using the video-magnetograph at the Big Bear Solar Observatory. This device (invented by yet another student of Leighton) measures the flux and polarity of the patches over a large area and presents a movie of the action. Martin observed patches of opposite polarity, about 2,000 km in size, approach each other at about a kilometer per second, merge, and *disappear*. This was occurring at the borders of active regions as well as in the magnetic network surrounding supergranules.

Martin called these events "cancellations" and tentatively suggested that she was seeing the actual *annihilation* of magnetic flux. This is a real physical possibility when oppositely directed fields of comparable strength contact each other. It would be a special case of magnetic *reconnection*, about which we'll have much more to say later.

Cornelis Zwaan was skeptical about magnetic annihilation and offered two alternative explanations for the cancellations. He reminded us that all flux is contained in loops, both small and large. He suggested Martin was observing the *submergence* of small loops. As a loop descends, she would see the feet of the loop in the photosphere draw closer, giving her the impression of a "cancellation" of flux. Or conceivably Martin might have seen *inverted* U-loops pulled up out of the photosphere. This would also give one the same impression of merging and disappearance of opposite polarities.

Can high-resolution observations decide whether either of these processes is active? Ideally one would like to follow the behavior of entire loops and not just their photospheric roots, but for the moment magnetic observations are limited to the photosphere.

Allan Title and his associates at the Lockheed Martin Palo Alto Advanced Technology Center are trying hard to learn how magnetic flux comes and goes. Title, a former Leighton student, has been at the forefront of solar experimentalists ever since he graduated from Cal Tech in 1962. He has pioneered high-resolution solar observations from space. His group built the magnetograph aboard the SOHO satellite that is now in orbit and also built the complex coronal instruments aboard the TRACE satellite, which was launched in April 1998.

Title has always stressed the *dynamics* of small fields. He prefers to use narrow-band tunable filters, rather than spectrographs, in order to make movies of the changes in tiny magnetic fields, over a sizable portion of the solar disk.

In 1996 Title and Tom Berger traveled to the Swedish Observatory in the Canary Islands to make a four-hour movie of "bright points." You may recall that Richard Muller, a French solar astronomer, demonstrated that these tiny features in the intergranular lanes, no larger than 200 km in diameter, are transient brightenings associated with clusters of magnetic flux tubes. The causes of these brightenings are still controversial but since they can be observed more easily than the magnetic fields of the tubes themselves, they make good tracers for the tubes.

Berger and Title observed that evolving granules constantly jostle the bright points, which jiggle frenetically at speeds between 0.5 and 5 km/s. They collide, merge into chains, and then separate again, all within a few minutes. As we shall see, this behavior is thought to be crucial for heating the corona.

The director of the Swedish Solar Observatory, Gören

Scharmer, has made movies of the actual magnetic fields in the photospheric network, at a higher spatial resolution, and over a larger area, than anybody else so far. He reports that patches of flux of all sizes (down to the 200 km limit of his telescope) emerge, collide, and "cancel" within the eight-minute lifetime of the granules. Clearly the next step will be to measure the flows within canceling patches to try to determine whether the plasma is moving down (as in a submerging loop) or up (as in a rising loop) or doing something different (as in an annihilating pair of loops).

Incidentally, the Swedish Solar Observatory is one of several large astronomical facilities perched on the rim of a high extinct volcano in the Canary Islands. The Nordic Telescope, the Dutch Solar Telescope, the French solar telescope THEMIS, and the British William Herschel telescope are all located there. The site was chosen after a search lasting two decades as the best within a reasonable distance of Europe. It has become a major center because of the superb image quality it offers day or night, at least between the dust storms that roll in from Africa. It's an example of the lengths (or distances!) astronomers will go to escape the bad observing conditions of home. Another example is the great cluster of international telescopes on Mauna Kea, in the Hawaiian Islands.

Meanwhile, Carolus Schrijver (another member of Title's group) and his colleagues have investigated how the continual emergence and disappearance of individual flux elements refreshes the photospheric magnetic network. They analyzed a twelve-hour time-series of magnetograms obtained in 1996 from SOHO (see endnote 7.5). The great virtue of these satellite observations is that they are completely uniform in quality, but their resolution is limited by the size of the onboard telescope to about 400 km.

They concluded that the network in the quiet sun, far from active regions, can be maintained by the EARs alone, even if one neglects the contribution of other small bipoles, such as the intranetwork fields. Their model predicts that all the flux in the network is replaced by new flux in three to five days, a remarkably short time. This study avoids the issue of whether the flux is annihilated or submerges. But it shows that if the flux does submerge, it is unlikely to reappear.

From these recent studies we get a picture of the network as a battlefield of competing flux loops. The loops pop up in the centers of the supergranule cells as EARs and their feet in the photosphere appear as bipolar magnetic patches. As convective motions separate the feet, the loop expands upward. But at the same time it is fragmented by granular motions into flux tubes and these tubes are swept into the downdrafts at the borders of the supergranules. There the tubes merge with others of the same sign or "cancel" those of opposite sign.

It is clear that a lot more remains to be discovered about the interactions of flux patches of very small size, down to the elemental flux tubes. Even more tantalizing is the interaction of the tubes with granular convective motions, that may excite magnetic oscillations or waves that can travel up into the higher atmosphere. As we shall see in the following chapters, some of this convective energy produces much of the complicated structure we see in the X-ray picture, Figure 7.1.

Chapter 8

THE MIDDLE KINGDOM: THE CHROMOSPHERE AND TRANSITION ZONE

O n July 18, 1851, hundreds of astronomers were stationed in Scandinavia and along the Baltic coast, waiting for a total eclipse of the sun to begin. Slowly the moon covered the sun's disk and the sky grew darker and darker. The crowds grew tense. Then, in the last few seconds before the moon covered the sun completely, and before the corona appeared, they saw a deep red crescent of light flash out at the edge of the sun. Those who were quick enough to get a good look, agreed that the outer edge of the crescent was jagged, like a range of mountains. They saw this "sierra" again for a few seconds at the end of the eclipse, just as the moon began to reveal the disk.

Sir Norman Lockyer, the eminent British astronomer, named this ruby red region the "chromosphere" (1869) and Charles Young, the discoverer of the coronal green line, captured its spectrum for the first time (1860). But Father Angelo Secchi, the director of the observatory of the Collegio Romano, was the first to get some idea of the strange shape of this region just above the photosphere. At the eclipse of 1860 in Spain, he made quick sketches of the sierra. He must have had extraordinary eyesight, for his drawings show long, thin needles of red light extending from the top of the sierra.

Further progress had to await new technology. George Ellery Hale, director of the Mt. Wilson Observatory, built the first spectroheliograph at the turn of the century. This instrument produces images of the solar disk in the light of individual spectrum lines, like the hydrogen alpha line or the calcium K line. Since the atmosphere is relatively opaque in these deep absorption lines, the light can only escape from a relatively high layer above the photosphere. So the spectroheliograph allowed Hale to look down on the top of the chromosphere.

Fig. 8.1 Spicules lie at
the borders of
supergranule cells, in
the magnetic network.
(Courtesy of R. Dunn)

Near active regions (Figure 6.1) the chromosphere resolves
into a tangle of loops, bright patches, and long, dark "fibrils."
In the quiet sun (Figure 8.1) we now can see that Secchi's
needles (the "spicules") are arrayed in long picket fences.
Bear in mind that these photographs show only the relatively cool plasma and that all the remaining space is filled
with tenuous plasma at coronal temperatures.

We now identify all this fine structure with magnetic fields
that shape and channel the plasma. The chromosphere and
the hotter regions within it, around it, and above it are the
extensions of the photospheric flux tubes. But the geometry of these two intertwined regions, and their relationship to the corona, is very complicated and has not been

unraveled to everyone's satisfaction.

Their spectra, or equivalently their temperatures, usually distinguish these regions from each other. In crude terms, the chromosphere consists of plasma between 5,000 and 50,000 kelvin, the transition zone between 50,000 and, say, 500,000 kelvin. If the sun were layered like an onion, a simple one-dimensional picture (like that of 6.11) could represent them very well. But Figures 6.1 and 8.1 show the true state of things, and challenge the astronomer to devise models for different types of structures. We'll examine several of these: spicules, "canopies," and transition zone threads. In addition, we'll talk about the coolest parts of the sun, the carbon monoxide clouds.

SPICULES

Walter Roberts, a dynamic Harvard graduate student, built the first coronagraph in the United States, following a design by the French astronomer Bernard Lyot, and used it in the mid-1940s to study the properties of Secchi's needles of red light. Roberts named them "spicules." He and his coworkers counted their numbers at different heights above the limb, measured their widths, and tracked their motions.

Roberts carried out his basic studies at Harvard's solar station near Climax, Colorado, at an altitude of 11,000 feet. For several years, he and his wife Jan were the only scientists living there full time. Jan recalls writing to the Betty Crocker Cake Company for advice in baking a cake. "How much yeast should one add to the batter at 11,000 feet?" she inquired. A formal letter on company stationery informed her that "nobody lives at 11,000 feet, you must have meant 1,100 feet." Roberts went on to establish the High Altitude Observatory, a major solar institution. His lifelong interest in solar influences on the earth's climate led him to press Congress to establish a National Center for Atmospheric Research at Boulder, Colorado. He was the Center's first director.

But I digress. Roberts and his colleagues learned that the "sierra" seen at the solar limb by the early eclipse watchers consists of thousands of spicules, overlapping along the line of sight, to form a uniform ring of light about 3,000 km in width. Much of this light is contained in a single powerful spectral line (hydrogen alpha, at 656.3 nm) in the red end of the spectrum. Above this ring the taller spicules can be resolved. Most lie below 6,000 km but a few extend to 8,000 or even 10,000 km. Their widths are so narrow (less than 1,000 km) that they can be photographed only during the best observing conditions.

Roberts found that a typical spicule rises above the photosphere at a speed of 20 to 30 km/s, reaches a maximum height between 6,000 and 10,000 km, and fades in place. A third to half of all spicules fall back to the surface. The whole event takes ten or fifteen minutes. Later on, spicule spectra indicated that spicule temperatures vary with height and time, and lie between about 6,000 and 15,000 kelvin.

In the quiet sun, far from active regions, most of the chromospheric plasma is contained in spicules. Moreover, as Figure 8.1 shows, spicules congregate at the boundaries of supergranule cells. As we have seen, the horizontal expansion of a supergranule sweeps magnetic flux into its boundaries to form a magnetic network. And we have also seen that the flux is constantly merging, canceling, and submerging in the network.

All these observations lead to the idea that a spicule is a kind of plasma jet that squirts up a flux tube. Most astronomers agree on that, but the source of its kinetic energy has been debated vigorously over the past thirty years and is still not settled.

Many theorists have tried to explain the life cycle of a spicule with a physical model. Eugene Parker was among the first, and has influenced the ideas of all that followed. In 1964, he imagined a vertical flux tube embedded among granules. At some moment the neighboring granules happen to converge on the deep roots of the tube and give it a hard pinch. The pressure inside the tube rises sharply and pushes a dense column of photospheric plasma up the tube. That column collides with the overlying static plasma in the tube and forms a shock wave. The wave runs up the tube at the speed of sound, compressing, heating, and propelling the plasma it meets. That, Parker suggested, is the spicule we observe above the photosphere. When the spicule expends its initial pulse of energy, it reaches its maximum height and either cools off and/or falls back.

Parker's effect is similar to stamping on a tube of toothpaste, except that the squirting paste remains cool, while the plasma is heated by compression.

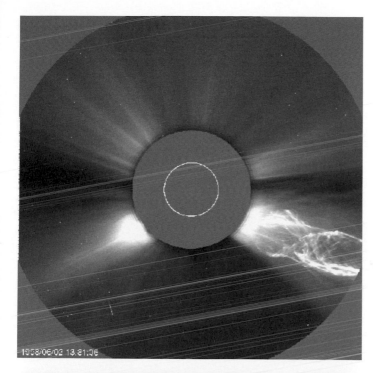

Figure 1-1. A coronal mass ejection in progress. The white circle shows the size of the sun. (Courtesy of SOHO/LASCO consortium. SOHO is a project of international cooperation between ESA and NASA).

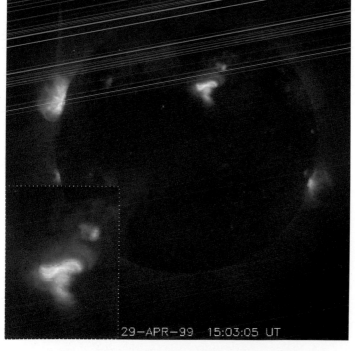

Figure 1-2. The sun in X-ray light. A single active region on the dark disk has an S shaped coronal magnetic field (shown enlarged in the inset) which is an indication that it may eject coronal mass within a few days. (Courtesy of the YOHKOH team. YOHKOH is a project of international cooperation between ISIS and NASA).

Figure 1-4. A sample of Fraunhofer absorption lines in the sun's visible spectrum. Each element emits unique patterns of lines that enable astronomers to determine its abundance in the sun.

Figure 1-5. The visible corona at the solar eclipse of July 11, 1991. Streamers are at their most beautiful near the maximum of the solar cycle. *(Courtesy of the High Altitude Observatory.)*

Fig. 5.1 Patches of the solar surface oscillating with a period of about 5 minutes. This image is a typical snapshot, showing the rising and falling patches. *(Courtesy of SOHO/ SOI/MDI consortium)*

MDI medium-*l* power spectrum

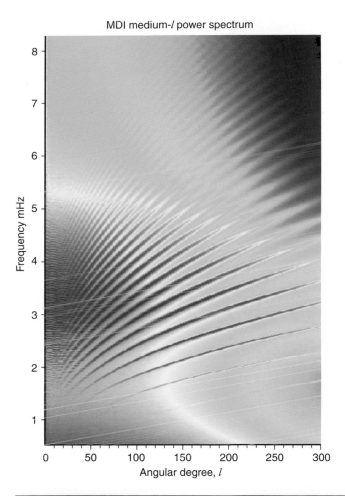

Fig. 5.2 A "diagnostic diagram," which arranges individual modes of oscillation according to their frequency (vertical axis) and horizontal wavelength (horizontal axis). All the modes with the same number of nodes along a solar radius $\underline{n}$ fall on the same discrete ridge. The ridge at lower right corresponds to $\underline{n}= 0$, the ridge at the upper left to $\underline{n} = 40$. (Courtesy of SOHO/SOI/MDI consortium)

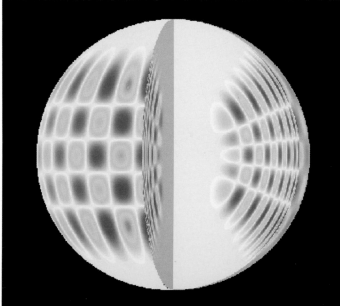

Fig. 5.6 An example of a three-dimensional standing wave pattern in the sun. Blue regions are rising, red regions are falling. (Courtesy of SOHO/ SOI/MDI consortium. SOHO is a project of international coop-eration between ESA and NASA.)

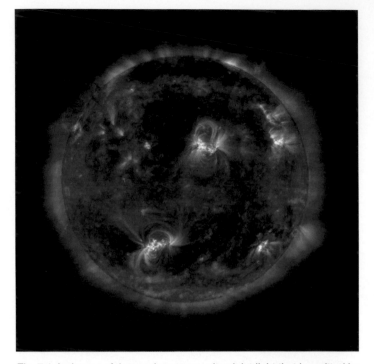

Fig. 7.1 An image of the sun in extreme ultraviolet light that is emitted by coronal plasma at a million kelvin. Coronal loops, shaped by the magnetic field, are visible above active regions. (Courtesy of SOHO/EIT consortium)

He I 584.34 A   O V 829.73 A   Mg IX 358.06 A

Fig. 8.5 The transition region in the quiet sun resides over the supergranule borders (the so-called network). Each panel shows the TR features visible at a characteristic temperature, ranging from 20,000 kelvin to over 1,000,000 kelvin. Because the features hardly change shape until coronal temperatures are reached, the TR is thought to exist in columns defined by the nearly vertical magnetic field. (Courtesy of SOHO/CDS consortium)

Fig. 8.6 The so-called "moss," a network of million-kelvin plasma lying at the feet of hot (3 million kelvin) coronal loops. (Courtesy of the TRACE consortium and the Lockheed-Martin Solar and Astrophysics Labs)

Helium (20,000°)

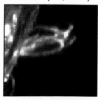

Oxygen (250,000°)

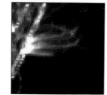

Neon (400,000°)

Calcium (630,000°)

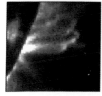

Magnesium (1,000,000°)

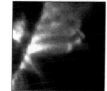

Iron (2,000,000°)

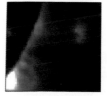

Fig. 9.5 An active region at the limb of the sun, as imaged by the Coronal Diagnostic Spectrometer (CDS) aboard the SOHO satellite. The different panels correspond to different spectral lines or plasma temperatures. (Courtesy of SOHO/CDS consortium)

SOHO/CDS 23-Mar-1998

Fig. 10.7 Coronal loops in an active region are anchored in sunspots (like the dark patch), pores, and intergranular lanes. (Courtesy of the TRACE team and the Lockheed-Martin Solar and Astrophysics Lab)

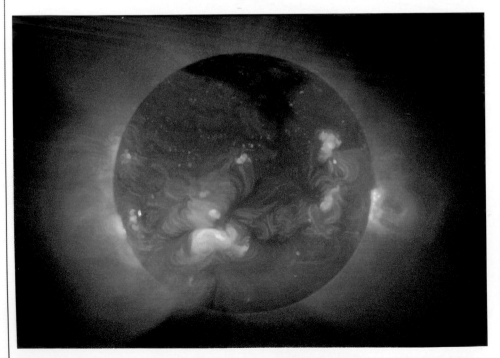

Fig. 11.2 An X-ray picture of the sun. The dark region at the north pole is a coronal hole and the bright loops are magnetic field lines made visible by the hot plasma they contain. (Courtesy of the YOHKOH consortium)

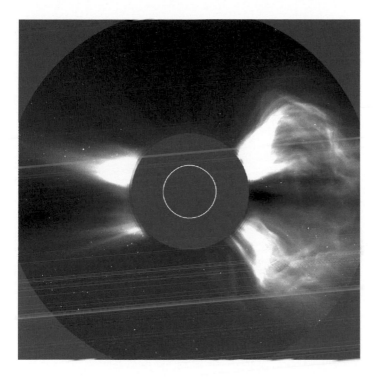

Fig.11.5 A huge coronal mass ejection in progress on November 6, 1997, as seen by the LASCO aboard the SOHO satellite. The white circle indicates the size of the solar disk. The two white blobs on the right are the "legs" of an enormous trans-equatorial loop that erupted as a CME. (Courtesy of SOHO/ LASCO consortium)

Fig. 11.7 Coronal loops resolved in a spectral line at 17.1 nm of Fe IX and Fe X, corresponding to a temperature of 1 million kelvin. (Courtesy of the TRACE team and the Lockheed-Martin Solar and Astrophysics Lab)

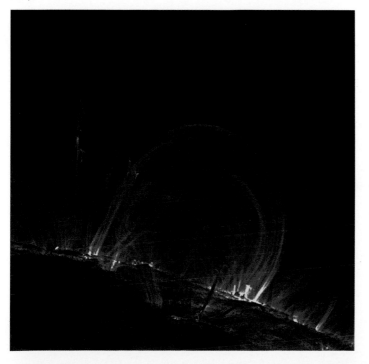

Fig.11.8 Small magnetic loops extend into the corona from magnetic bipoles in the photosphere, forming the "magnetic carpet" in the quiet sun. The bright patches were imaged in the 19.5 nm line of Fe XII by the EIT on SOHO, and show where the low corona is heated. (Courtesy of SOHO/MDI/ EIT/CDS consortia)

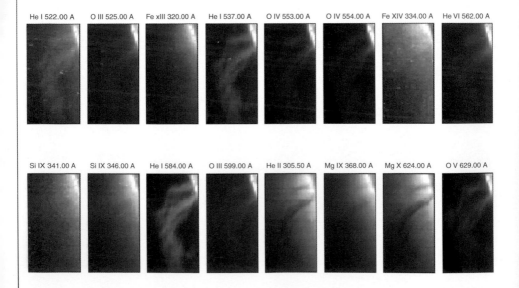

Fig.12.3 A single prominence as imaged in different spectral lines. Each line corresponds to a different plasma temperature (e.g., 20,000 K for He I 584 A; 200,000 K for O V 629 A; 2,000,000 K for Fe XIII 320 A). (Courtesy of SOHO/CDS consortium)

Parker's basic idea is to inject some mechanical energy at the base of a rigid flux tube and allow that energy to propel the spicule. Other theorists proposed other forms of energy or stressed other kinds of impulsive forces. The tension in magnetic field lines might sling the spicule upward. Or magnetic flux of opposite polarities in the network might simply annihilate each other, releasing heat to eject a spicule. Or heat from the coronal end of the flux tube might flow down and "evaporate" the dense plasma of the photosphere to create a spicule. Whatever their preferred mechanisms, such models are alike in accelerating the spicule briefly and allowing it to glide to its maximum height, like a ball in the air.

The problem with such models is that they require unrealistic speeds at the base of a spicule. To reach a height of 10,000 km the spicule would have to start with a speed of 74 km/s and such high speeds are never observed. In addition, the models usually predict too high an initial temperature and too much expansion of the width of the spicule.

One way out of these problems might be to use magnetic forces to lift the plasma continuously instead of impulsively. A train of hydromagnetic waves of the right kind might be able to do this. And so theorists have investigated various waves. The most appealing type is the Alfvèn wave, named for Hannes Alfvèn, who was later awarded the Nobel Prize for his pioneering work. Alfvèn was a Swedish astrophysicist whose uncle Hugo was a famous composer. Hannes, under the name of Olaf Johannesson, wrote a political satire titled "The Great Computer," about a computer that takes over the world. He may be the only Nobelist who wrote a science fiction novel.

Figure 8.2 shows three kinds of Alfvèn waves that are produced by shaking, squeezing, or twisting the bottom of a flux tube. In each type the distortion of the field lines travels up the

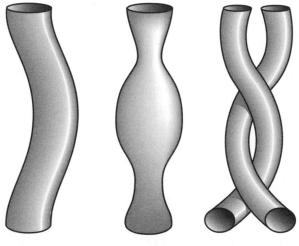

Fig. 8.2 Three types of Alfvèn waves, in which field lines are bent, bulged, or twisted.

Bend   Bulge   Twist

tube at a speed that depends on the field strength and plasma density. The stronger the field and thinner the plasma, the faster the wave travels. A typical speed in chromospheric conditions is about 100 km/s.

Alfvèn waves are transverse waves, like those on a string. Although the bulge in the field travels along the tube, the field lines themselves move only sideways. The plasma, however, can be forced to slide up the tube if the magnetic pressure of the wave is strong enough. If the plasma speed reaches the local speed of sound, which is determined by the local temperature, the plasma will shock (like a sonic boom) and convert its kinetic energy to heat.

Granules are the most likely movers and shakers of flux tubes. As we have seen, the flux tubes are swept into the intergranular lanes by the horizontal granule motions. In the lanes the roots of the tubes might be buffeted, pinched, or twisted by fast plasma motions. Numerical simulations suggest the motions could be quite brisk. But all this is con-jecture so far—the lanes are so narrow that no detailed ob-servations of such motions have been made. Nobody has reported seeing Alfvèn waves being born in the lanes, much less anything about their properties. So for the moment, theorists are free to guess.

Several theorists have toyed with Alfvèn waves in an at-tempt to explain spicules. As an example, we pick Joe Hollweg at the University of New Hampshire, who has made the most detailed studies. In 1982 he and some col-leagues made pioneering numerical simulations of waves on a flux tube, following their evolution even beyond the time that they shocked (see endnote 8.1).

But Hollweg's spicule is much too cold to match observa-tions. The waves in his model lift the plasma effectively, but don't heat it sufficiently. He suggested that some addi-tional wave dissipation mechanism, including "ion-neutral friction," might help. Recently Gerhard Haerendel, a Ger-man plasma physicist, rediscovered Hollweg's suggestion and worked out its consequences (endnote 8.2 describes this mechanism).

He found that he can produce a respectable spicule if his Alfvèn waves build up to a transverse speed of 40 km/s. From previous work by Hollweg and others, this does not seem impossible. But Haerendel also needs wave periods in a very narrow range, between 3 and 10 seconds and nobody knows whether these are likely.

So we might say that we are not home yet.

THE SUN'S SIBERIA

Somewhere in the solar atmosphere, between the photosphere and the corona, the plasma temperature must drop to its minimum. Common sense would say that the minimum lies just above the photosphere. And indeed the influential temperature models of the Harvard group, which were derived from massive quantities of observations, show just that: a minimum temperature of 4,200 to 4,700 kelvin at a height of 500 or 600 km above the photosphere (Figure 6.10 is an early example). In separate models for the quiet sun, bright network, and active regions, the details of the temperature region differ somewhat, but not much. By the mid-1980s a comfortable consensus was growing that the temperature profile of the low chromosphere was understood.

And yet, lurking on the sidelines of the discussion, was a disturbing set of data that just didn't fit this cozy picture. Tom Ayres, a young astrophysicist at Harvard, claimed in 1986 that somewhere in the solar atmosphere the temperature drops as low as 3,000 kelvin. He based his claim on spectra of the carbon monoxide (CO) molecule that he and his colleagues had obtained (see endnote 8.3).

When Ayres and company analyzed their data they found much to their surprise that the core brightness of the CO lines corresponded to a temperature as low as 3,000 kelvin while the K line profiles in the same location on the disk corresponded to 4,000 kelvin or higher. To account for this discrepancy, they concluded that each patch of the sun contains a mixture of objects, some cool and some hot, that were not spatially resolved in their observations. The hot component, they suggested, originates in flux tubes within the magnetic network and fills as little as 5% of the solar surface while cold clouds (over the cell interiors) fill the remaining 95%.

Their claims were met with much skepticism. After all the work that had gone into the Harvard models, and all the corroborating observations that followed, some astronomers could not readily accept that their ideas had to change. In particular, Grant Athay, the acknowledged expert on chromospheric modeling, was most vocal in his disbelief. He and Ken Dere of the Naval Research Labs used ultraviolet spectra of carbon, oxygen, and iron atoms to estimate the fraction of the solar surface covered by the hot component (endnote 8.4).

Fig.8.3 A portion of the infrared spectrum of carbon monoxide, showing the typical band structure. (Courtesy of T. Ayres)

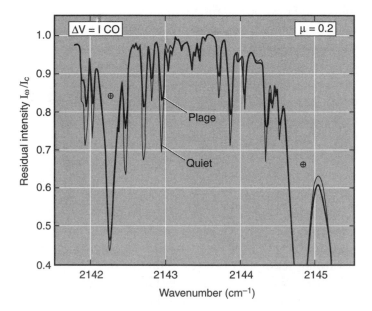

Athay and Dere concluded that 90% of the area chromosphere must be hot, in outright contradiction to Ayres and his colleagues! Perhaps, they suggested, the cool component is nothing more than the intergranular lanes, cool and very small in fractional area. It could not conceivably lie in the classic chromosphere.

In order to convince Athay and other skeptics, Ayres had to show that the cold CO clouds actually reside in the chromosphere and that they must cover a large fraction of the solar surface. In addition, they had to locate just where these cold clouds were floating.

A breakthrough in the controversy was made in 1994, when Ayres, Bill Livingston, and Sami Solanki detected CO in emission above the solar limb. They were able to measure the position of the emission and show that indeed it lies in the chromosphere at a height of a few hundred kilometers. Then, later in 1994, they made additional observations on the solar disk, including the K line, and constructed a two-dimensional model that resolved the apparent contradictions among different temperature sensors.

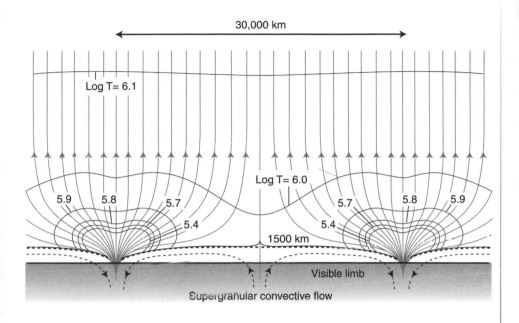

Their picture of the quiet chromosphere resembles Figure 8.4. Above the network the flux tubes spread out to fill all the available space. But in the lower atmosphere, the field lines are virtually horizontal and form a so-called "canopy" over the supergranule interiors. Within the flux tubes and their extensions, the temperature rises steeply at a height of about 1,000 km. It is here that the K line originates, and indicates that the temperature is hot. Under the canopy, the supergranule cells are nearly free of the magnetic field (if we ignore the weak intranetwork fields), and the plasma is cooler and emits in the CO bands. Now one can see how each party to the controversy could conclude that most of the area covered by the chromosphere is at a very different temperature. The K line region fills all the space above a height of about 2,000 km and the CO lines fill nearly all the space *below* 2,000 km.

Fig. 8.4 A magnetic field that emerges from the borders of supergranule cells in the so-called "coarse network." The field fans out into the overlying corona. Over the center of the cell lies a horizontal magnetic canopy. The numbers in the figure denote the logarithm of the plasma temperature. (Courtesy of A. Gabriel)

Why was this issue important? What difference does a few hundred kelvin make in a narrow layer of the sun? One answer concerns the nonthermal energy that must be supplied at each height to maintain the temperature of the atmosphere. In Athay's or Avrett's models, the radiative losses rise rapidly just above the height of temperature minimum. A few hundred kelvin more or less in the temperature region could make a big change in the losses and therefore in the necessary energy supply. And that bears on the mechanisms that might be doing the heating.

This story has a couple of morals. First, apparent contradictions can arise when the multi-dimensional form of the solar atmosphere is not fully appreciated. Secondly, like many scientists, solar astronomers are reluctant to break a well-established paradigm that explains all their previous experience. Thomas Kuhn has described this process in his book, "The Structure of Scientific Revolutions."

THE TRANSITION REGION

Plasmas in the sun's atmosphere, with temperatures between, say, 50,000 and 500,000 kelvin, comprise the transition region (TR). If as we said earlier, the sun were layered like an onion, the transition region would lie between the chromosphere and the corona and there would be no uncertainty about it. But we've seen that the chromosphere is not a smooth layer. In the quiet sun, the chromosphere is brightest above the supergranule borders (the "network") and rises into the corona as needle-shaped spicules. In active regions, there are hot loops of all sizes. Where in this picture is the transition region plasma?

The best guess is that the transition region is an *interface*. It appears wherever the temperature jumps abruptly from chromospheric to coronal values. Unfortunately, that definition covers a lot of territory. Astronomers have been trying to pin down the TR's location and properties for some thirty years now and still don't have all the answers.

The region radiates only in the extreme ultraviolet at short wavelengths that don't penetrate the earth's atmosphere. So astronomers had to wait for the space age to study the region. After some preliminary work with rockets, the real breakthrough came during the Skylab Mission in 1973. Skylab produced *images* of the transition zone plasma, in spectral lines that form at different temperatures. Details larger than about 3,500 km could be resolved.

Figure 8.5, made with the Coronal Diagnostic Spectrometer (CDS) aboard SOHO, shows much the same kinds of images. The blobs we see here are bright elements over the chromospheric network. They look much the same, regardless of which spectral line you choose, until you reach a characteristic temperature of about a million kelvin, whereupon the network expands to fill the whole corona.

Alan Gabriel, a physicist at the Culham Laboratory, offered a reasonable model for the shape of the TR, based on Skylab

He I 584.34 A   O V 829.73 A   Mg IX 358.06 A

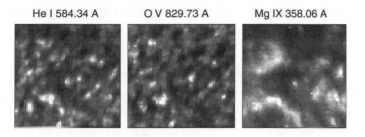

Fig. 8.5 The transition region in the quiet sun resides over the supergranule borders (the so-called network). Each panel shows the TR features visible at a characteristic temperature, ranging from 20,000 kelvin to over 1,000,000 kelvin. Because the features hardly change shape until coronal temperatures are reached, the TR is thought to exist in columns defined by the nearly vertical magnetic field. Also see color supplement. (Courtesy of SOHO/CDS consortium)

observations (Figure 8.4). In a unipolar region of the quiet sun the magnetic field lines are anchored in the network and spread out like the branches of a tree above the photosphere. Gabriel suggested that the TR plasma occupies the funnel formed by the field, with hotter plasma lying above cooler. The onion layer model has been replaced with a columnar model.

This nice, simple picture began to change in the early 1980s, when a team at the Naval Research Lab began taking TR spectra and images with an instrument they called the High Resolution Spectrograph and Telescope (HRTS). This rocket-borne instrument could resolve features at least five to ten times smaller than those aboard Skylab, that is, 300 to 700 km. The group expected to be able to trace chromospheric features into the TR very accurately with their superior spatial resolution. But they were surprised to find that the physical connections between the two regions were not at all obvious.

As an example, take spicules. In one flight, HRTS imaged spicules at the limb in C IV (present at about 100,000 kelvin) and simultaneously in neutral hydrogen, at 10,000 kelvin. The pictures show that most C IV spicules do not extend from cooler hydrogen spicules. Perhaps another whole population of hot spicules exists in the TR, independent of the chromospheric spicules.

In addition, the HRTS flights produced troubling evidence that the true dimensions of TR features were unresolved, even at 300 km. The evidence came from attempts to determine the plasma density in the TR from the spectra. Different methods of estimating the density gave answers that differed by a factor of ten. If one believes that the estimates are both correct, then the only way to reconcile them is to conclude that the spectra are emitted from very thin, very dense filamentary threads. They may be as narrow as 50 km, and fill only 1% of the available volume above the chromosphere. The NRL team suggested that these threads could be aligned along magnetic field lines, but other possibilities exist too.

HRTS also revealed the violent nature of the TR (endnote 8.5).

The long series of HRTS observations left astronomers wondering just where the unresolved TR plasma is located within the gross structures they can observe. One NRL astronomer has proposed that the unresolved TR structures are disconnected somehow from both the chromosphere and corona. A trio of scientists suggested the TR emission arises in inconspicuous, low-lying cool loops. Still others proposed that the TR plasma lies in sheaths around spicules. In principle, anyplace the temperature jumps from chromospheric to coronal values is a potential location. Evidently, better observations were needed to sort out the possibilities. With the launch of two satellite observatories, SOHO and TRACE, astronomers have another opportunity.

These observatories are producing a flood of new results, particularly concerning the corona, as we'll see. In May 1998 a team of scientists pinned down the site of transition zone plasma in TRACE images of active regions (Figure 8.6). They called "moss." At a resolution of 700 km, it consists of an irregular bright network that is most visible in plasma

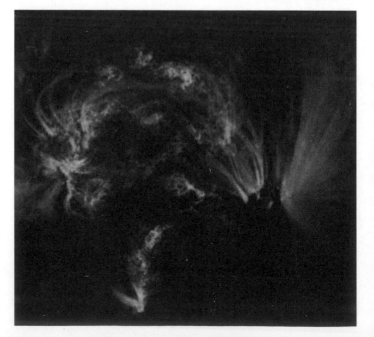

Fig. 8.6 The so-called "moss," a network of million-kelvin plasma lying at the feet of hot (3 million kelvin) coronal loops. Also see color supplement. (Courtesy of the TRACE consortium and the Lockheed-Martin Solar and Astrophysics Labs)

between 600,000 and 1,000,000 kelvin. The moss only appears in active regions that have very hot coronal loops (3 to 5 million kelvin) and lies in a layer 2,000 km thick at a height of 3,000 km above the photosphere. The team concluded that the moss represents the upper transition region in the feet of hot coronal loops and may radiate heat that is conducted down the legs of the loops.

The moss is extremely dynamic. It displays fast plasma flows, waves, and rapid changes in brightness. All this activity, in even the "quiet" sun, is related to heating, as we shall see later.

Chapter 9

ACTIVE REGIONS

alileo would have been astounded to see an X-ray image of the sun, like Figure 7.1. The smooth face of the sun he knew so well is covered with brilliant bundles of hot plasma. These are the sun's active regions, sites of intense magnetic fields and violent change. They are complex structures, composed of nested loops of magnetic flux that seem to compete actively for living space, like plants in a jungle. To solar physicists they are beautiful, awesome, and puzzling, as they would be to Galileo.

Some of the most important questions about them still have no definite answers. How are they heated? How does all their magnetic flux disappear? What do they look like just under the surface? In this chapter we'll explore their complicated metabolism and survey some partial explanations.

THE BIRTH AND DEATH OF A REGION

Like a baby in a womb, an active region begins life deep within the sun, somewhere near the base of the convection zone. At the present time we can't observe its birth or its early life as it struggles toward the surface. These are matters that theorists are debating, and that helioseismology may eventually clarify. We can, however, observe the emergence of a region and from that infer something about its early development. In chapter 13 we'll discuss the birth pangs again but for now we have the following sketch—a scenario—of how it might happen.

Some 200,000 km below the photosphere, the solar dynamo produces horizontal ropes of magnetic flux that girdle the sun, parallel to its equator. The ropes are lighter than their field-free surroundings and like hot air balloons they tend to rise. In most places, the slow convective motions are strong enough to keep them submerged. But somewhere along the length of a rope, a hairpin kink may bulge upward and begin to lift toward the surface (Figure 9.1). Astronomers call such a kink an "omega loop" from its resemblance to the Greek letter Ω.

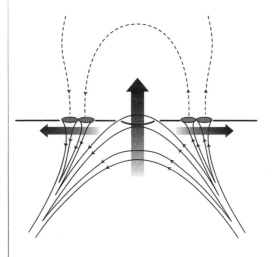

Fig. 9.1 An "omega loop" that reaches the photosphere after rising through the convection zone. The dotted line shows the extension of the loop into the corona. Where the different strands of the loop cross the top of the photosphere, pores and sunspots become visible.
(Courtesy of C. Zwaan)

Theorists tell us that original rope must have a field strength of at least 10,000 gauss to resist being shredded by the convective motions. But to resist the Coriolis force, which acts to wind up the rising loop, its initial field strength may have to be as large as 60,000 gauss. Such powerful fields are not observed at the surface, so at each point in its journey the loop must adjust its magnetic pressure to match the external plasma pressure. It expands laterally as it rises, and frays into a loose bundle of thinner loops. When the loops arrive, after a journey of several months, their field strengths are down to a few hundred gauss.

When the bundle of loops breaks through the surface (see Figure 9.2), its legs in the photosphere appear as a spattering of small magnetic bipoles. The largest of these will be the main pair of sunspots of the new region, the smaller ones are so-called "pores"—tiny sunspots without penumbras. The bipoles arrive with random orientations but soon align themselves (somehow!) in roughly the east-west direction.

The magnetic field near the larger spots is strong enough to inhibit convective heat flow from the neighborhood, so the spots cool, shrink, and darken. As they shrink in area, their flux is concentrated and their field strengths rise. At the same time, strong downdrafts develop in the young spots and they suck in pores of the same polarity. In this way they grow in area. In effect they eat their neighbors.

Over the next day or two the areas of opposite polarity in the photosphere ("plages") drift apart at a speed of about a kilometer per second. Simultaneously, the upper parts of the loops continue to rise and excess plasma in them drains down the legs of the loops, into the photosphere.

When the loops reach a height of a few thousand kilometers, they become visible in the hydrogen H alpha line as a system of "arch filaments," that connect the areas of oppo-

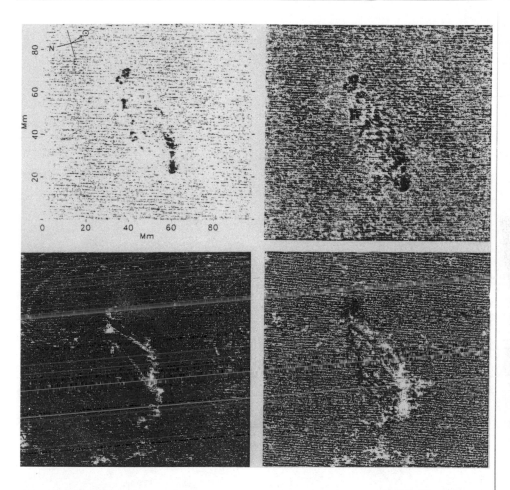

site polarity in the photosphere (see Figure 9.2 and endnote 9.1).

Slow plasma motions in the photosphere tend to tangle the otherwise simple omega loops and create havoc. As long as new flux arrives from below, the plasma in the legs of loops flares and surges dramatically. As we shall see in the next chapter, this constant low-level activity is a clue to the heating that goes on.

After a week or two, a mature region will extend far into the corona as a nest of loops (Figure 9.3). But once magnetic flux stops erupting, the region begins to decay. Like a large corporation, a region either continues to grow or dies!

A region decays by losing flux as it spreads over a larger and larger area. The spots fade out first, fraying into a looser bundle of flux and spreading outward. Supergranule cells nibble at the region's borders, spreading flux far over the solar surface, as we have discussed before. The processes

Fig. 9.2 Observations of an emerging active region. The top left panel shows pores and spots appearing in the photosphere, the top right panel shows the region just above the photosphere. At the lower left, we see magnetic loops, outlined in plasma that connect opposite magnetic polarities. The last panel shows magnetic field strength in the photosphere: white is positive, black is negative. (Courtesy of L. Strous)

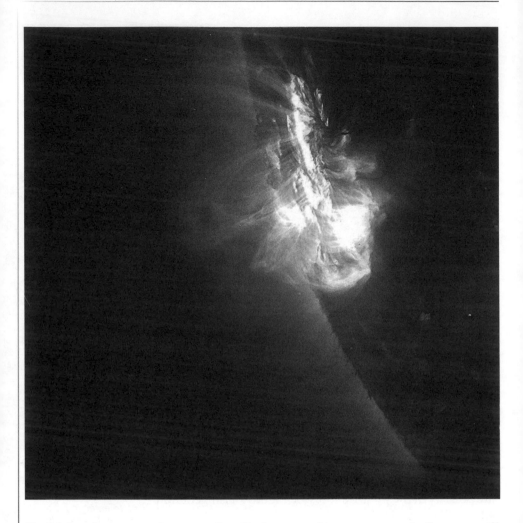

Fig. 9.3 An active region at the limb of the sun. The magnetic loops are outlined by plasmas with temperatures between 1 and 3 million kelvin. (Courtesy of the TRACE team and the Lockheed-Martin Solar and Astrophysics Lab)

that actually eliminate or destroy magnetic flux are still rather uncertain, however. Karen Harvey and associates found recently that most flux probably submerges, while some may cancel with flux of opposite polarity in the neighborhood. Whether true annihilation of flux occurs during the decay phase is still a question. And how much of the original flux remains, alive and well in the quiet sun, is also debated.

THE THERMAL STATE

When we look at an X-ray picture of an active region we get the impression that it consists of discrete loops of plasma and in a sense it does. But bear in mind that the whole coronal volume of a region is filled with a continuous magnetic field. Only some field lines or flux tubes within the field are filled with plasma, however, and it is these that show up in a picture. Recent observations from TRACE show that the hottest loops lie low in the core of the region and are surrounded by high, cooler loops. The loops are very thin; 700 km wide or less.

Why only certain tubes are filled is a big question. The best guess is that only some tubes are heated sufficiently (by processes we'll discuss later) to supply the radiation from a hot plasma. The heating is sporadic and turns on and off at different places and at different times, so that different loops brighten and fade. The gross structure of a region changes only very slowly, however, over weeks or months.

If we ignore all these complications and just take a snapshot of an active region we can determine how much plasma exists at any temperature. Figure 9.4 shows the result. The first thing to notice is that plasma temperatures cover a huge range—from 10,000 to 5 million degrees and sometimes higher. But most of the plasma lies in two broad regimes: less than or more than 200,000 kelvin. In fact, most of the plasma lies between 2 and 3 million kelvin with a small fraction (about 5%) as hot as 4 to 5 million. This general picture has been confirmed by observations from YOHKOH, TRACE, and SOHO (endnote 9.2).

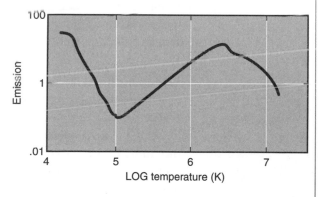

Fig. 9.4 The relative amounts of active region emission at different temperatures. Most of the mass is either at coronal or transition zone temperatures.

To make some sense out of such observations from Skylab, astronomers have tried a variety of methods. Bob Rosner, Bill Tucker, and Guiseppe Vaiana were the first to look for a so-called "scaling law" that relates several observable properties in single loops. Back in 1978, they used an argument based on the energy balance of a loop. In equilibrium, all the energy deposited in a loop by the unknown heating mechanism must escape as radiation, either from the sides of the loop or from the chromosphere and transition region in the legs. Heat conduction would carry excess energy from the top to the legs (see endnote 9.3).

Since Skylab, progress has been blocked by the fact that heat conduction along a coronal loop is so efficient that it practically wipes out any signature of a preferred heating

Fig. 9.5 An active region at the limb of the sun, as imaged by the Coronal Diagnostic Spectrometer (CDS) aboard the SOHO satellite. The different panels correspond to different spectral lines or plasma temperatures. Also see color supplement. (Courtesy of SOHO/CDS consortium)

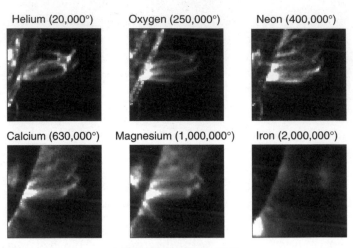

mechanism. Very recently, however, Eric Priest, a brilliant theorist at Saint Andrews University, and his colleagues have found some important clues. We'll discuss them in chapter 11.

Many astronomers have tried to account for the peculiar distribution of plasma temperatures in loops, as shown in Figure 9.4. A simple model of a loop, like that postulated by Rosner and associates, can explain the high-temperature branch of the diagram. The temperature varies from hottest at the top to coolest at the feet of the legs. But the low-temperature branch (at transition zone temperatures between, say, 50,000 and 200,000 kelvin) has remained a puzzle.

THE MAGNETIC FIELD

X-ray pictures reveal beautiful bipolar arrangements of coronal loops in active regions that resemble the "iron-filings" diagrams of magnets in high-school texts. There is absolutely no question that the loops outline magnetic fields. But astronomers have been thoroughly frustrated in trying to measure the field strengths.

The obvious technique, that has been so successful in the photosphere, is to measure the Zeeman splitting of coronal spectral lines. Unfortunately the splitting is quite small compared to the intrinsic width of the lines. So this direct method fails.

Until recently the best one could do was to measure the vector magnetic field in the photosphere of an active region and then, with some assumptions, *compute* the field configuration in the corona. With powerful computers this technique has become highly developed. Figure 9.6 is an example of the field in a real active region, taken from the work of Zoran Mikic and his colleagues at the University of California at Riverside. Several other groups have obtained similar results (endnote 9.4)

Potential field   Force-free field

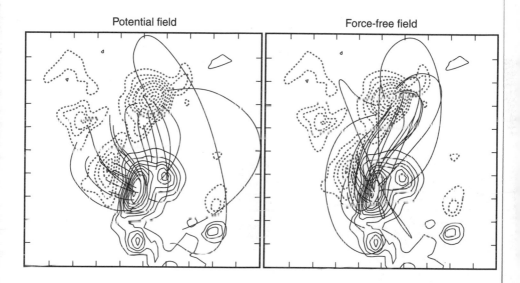

As a check on their calculations they compared the shapes of their computed field lines with chromospheric and X-ray images of the same active region. The match is fairly good, giving us some confidence in their methods.

Although such agreement is encouraging, most astronomers would prefer to see some independent confirmation. Several groups of astronomers have obliged, using a combination of radio and ultraviolet observations. As an example, consider the work of Jeffrey Brosius and colleagues at the NASA Goddard Space Flight Center, collaborating with Stephen White of the University of Maryland.

This group made maps of the polarized radio emission at wavelengths of 6 and 20 centimeters at the Very Large Array, a huge interferometer in New Mexico. These two wavelengths are well suited to probe the lower corona. Simultaneously, spectra and images of the region were obtained during a rocket flight of the Solar EUV Rocket Telescope and Spectrograph (SERTS).

The microwave emission depends on the temperature, the

Fig. 9.6 Calculated magnetic fields in the corona above an active region. The force-free model includes electric currents, the potential field model does not. (Courtesy of Z. Mikic in SPO Workshop "AR: Theory and Obs")

plasma density, and the magnetic field over each point in the active region. From the extreme ultraviolet spectra (30–40 nm) they extracted information on the plasma temperature and density. Using these data they were able to determine the height variation of magnetic field strength. Fields of several hundred gauss are not unusual in the corona above active regions.

SUNSPOTS

Sunspots have been studied intensively ever since Hale's discovery of their basic magnetic character, in the early 1900s, and yet we are still far from a thorough understanding of their structure and evolution. We don't even have a good explanation for the fact that they have an umbra and a penumbra. How are they heated? How do they die? What do they look like underneath the photosphere?

One would think that better observations, with sharper detail, would help to answer the kinds of questions everyone raises, but as observations have improved, a flock of new puzzles has appeared. Any recent image of a "simple" sunspot (like Figure 9.7) shows more complication than anybody can deal with. And this is only a snapshot! If you could see a movie of this spot you would be staggered at the variety of its intricate motions.

So there are no final answers—luckily for the astronomer—but we can survey a bit of what is known and what is sought.

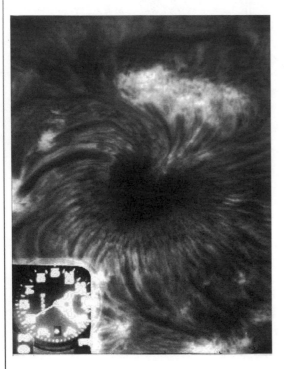

Fig. 9.7 An isolated sunspot. The penumbral filaments are beautifully resolved. (Courtesy of Big Bear Solar Observatory)

A PRIMER

Sunspots come in all sizes, from tiny "pores" less than 300 km in diameter, lasting a few hours, to giants 100,000 km across that can last for months. Earlier we mentioned that mature sunspots begin life as pairs of pores, and that they grow by accumulation of other pores and even smaller neighboring magnetic elements.

After a period of growth, a spot has a dark umbra and a dim surrounding penumbra, whose width is about the same as the umbral radius. A pore, on the other hand, has no penumbra. At high resolution the penumbra resolves into alternating bright and dark fibrils, about which we'll have much more to say.

The simplest picture of the magnetic field in a nice, round sunspot is shown in Figure 9.8. The magnetic field is nearly

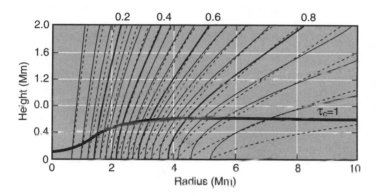

Fig. 9.8 A model of the magnetic field lines in a sunspot. Half of a cylindrical spot is shown, the axis of the spot is at radius zero. The heavy dark line shows the level to which one sees in white light. (Courtesy of V. Pizzo)

vertical in the umbra and fans out to lie nearly horizontally in the penumbra. The reason for this is simple. A bundle of magnetic flux exerts pressure on its surroundings—it wants to expand. In the photosphere the high plasma pressure can balance this magnetic pressure and constrain the flux of a sunspot to a tight circle. But the plasma pressure falls off rapidly above the photosphere and so the field is free to expand to fill the space above it. The field lines fan out rapidly and in the penumbra they are practically horizontal. A mature spot must look something like this within a few days of its emergence. Remember, however, that a spot is just the end of a loop, and its field lines must return to the photosphere somewhere.

Umbral fields reach 2,000 to 3,000 gauss, regardless of the size of the umbra. (Recall that the earth's field strength is only half a gauss.) The field strength decreases smoothly through the penumbra, dropping to a third or half of the umbral field just inside the sharp boundary with the surrounding photosphere and then (presto!) vanishing just outside the boundary.

In white light, the umbra and penumbra are approximately 20% and 75% as bright as the photosphere, respectively.

That means the umbra is about 2,000 kelvin cooler than the 6,000 kelvin photosphere. Recently, Per Maltby and his Norwegian colleagues discovered that the brightness ratio of the umbra and photosphere varies in a sawtooth fashion during the eleven-year cycle, at least in the deepest portions of the visible photosphere. (Another puzzle!)

Generations of astronomers have measured the brightness of the different parts of sunspots, and at all wavelengths. From such photometric data, it's possible to construct model atmospheres, similar to those for the quiet sun. In fact, the same people (e.g., Gene Avrett and company) often engage in this pursuit. They assume hydrostatic equilibrium prevails in the spot and determine the vertical temperature and density distributions that successfully predict the observations. Figure 9.9 is an example of such a model for the umbra of a typical spot. The spot has a broader temperature mini-

Fig. 9.9 Several empirical temperature profiles of a sunspot. Different investigators derive slightly different curves, but all show a deep, broad temperature minimum. (Courtesy of Kluwer Academic Publications)

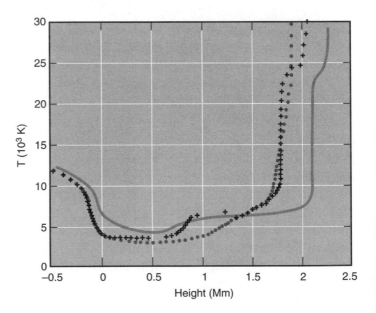

mum and a steeper chromospheric rise than the quiet photosphere.

In 1908, John Evershed, observing at the Kodaikanal Observatory in India, discovered systematic plasma flows in sunspots. In the photosphere the plasma streams radially outward from the umbra, through the penumbra, and as much as 10,000 km beyond, at speeds between 0.5 and 5 km/s. In the chromosphere of the spot, the flow reverses into the umbra, at speeds of 10 to 20 km/s.

SPOTS AS THERMAL PLUGS

At the maximum of the sunspot cycle less than 1% of the photosphere is covered with spots. Nevertheless, since a spot is dark, some of the radiation that would normally emerge at its location is missing. Is the sun therefore somewhat darker at sunspot maximum? If so, what has happened to the missing energy? These simple questions have puzzled solar astronomers for decades. Some satisfying answers have finally been offered.

The energy that escapes the photosphere is carried there by subsurface convection, as we described in chapter 4. But the strong magnetic fields in and around sunspots inhibit convective motions, at least down to some depth, and as a result the spots are dark. We might imagine that a spot acts as a thermal plug in the photosphere that blocks heat from surfacing. Does the heat just bypass this obstruction and emerge anyway, say, in a bright ring around a spot? That was the first idea that was proposed but it turned out to be wrong. Astronomers searched in vain, but there were no bright rings to be found.

So Gene Parker proposed that the spot doesn't *deflect* heat from below but *converts* it instead to Alfvèn waves. These waves would carry energy up the field lines and dump it into the chromosphere, where it could be radiated away. A thorough search was made and some evidence for Alfvèn waves was found, but if they exist, they are too weak to account for the missing sunspot energy.

And yet the presence of a spot at the surface is unlikely to affect the production of energy in the core. So a constant flow of energy must continue toward the surface despite the presence of sunspots. If the energy cannot escape immediately around or above the spot, it must escape someplace else and possibly at some later time. How is this possible?

Henk Spruit, among others, has thought deeply about the matter. He is a Dutch astrophysicist and a former student of Cornelus Zwaan. He pointed out two important properties of the solar convection zone that are relevant to the sunspot problem. First, turbulent convection is extremely efficient in transporting and redistributing heat. Any excess heat, such as that blocked by a new sunspot, could be redistributed throughout the convection zone in about a year. Secondly, the zone, being so massive, has an enormous capacity for storing heat. If the thermonuclear sources of energy in the solar core were turned off, the convection zone would not cool appreciably for approximately two hundred thousand years.

Put these two properties together and you have an explanation for the missing sunspot energy. It is simply redistributed throughout the convection zone and stored there. The zone's heat capacity is so large that the blocked heat hardly changes the temperature distribution in the zone or at the photosphere. Over a period of two hundred thousand years, the excess heat is slowly released, all over the solar surface. Therefore one should not expect to find a bright ring around each sunspot.

Since the convection zone redistributes energy relatively quickly but then releases it slowly, the appearance of a new spot will actually change the amount of sunlight received by the earth. The Active Cavity Radiometer Intensity Monitor (ACRIM) aboard the Solar Maximum Mission demonstrated this effect dramatically in 1980. ACRIM was capable of measuring the sun's energy output with a precision of a few parts in a million. When a large new spot rotated on to the visible solar disk, the measured output dropped by as much as 0.2% and rose again when the spot rotated off the disk. This proved that no bright ring existed around the spot, for such a ring would compensate for the dark spot and the ACRIM would never notice the presence of the spot.

THE DEVIL IS IN THE DETAILS

As observations of sunspots have improved, more and more of their intricate internal structure has been revealed to challenge astronomers.

In the umbra, bright "dots" are seen. They are so tiny (as small as 200 km) that their properties are hard to measure, and remain debatable. They are slightly fainter than the normal photosphere, and persist for tens of minutes. Some astronomers have suggested that they are convective cells that manage to poke up through the umbra from below. But no rising motions have been detected in them, nor any fluctuations in the umbral magnetic field. They remain something of a mystery.

The real complications begin in the penumbra, however. The first sharp pictures of sunspots, obtained from a high-altitude balloon in 1960 (the "Stratoscope"), showed that the penumbra is composed of radial alternating bright and dark "fibrils," as long as 5,000 km and only 300 km wide. Each fibril lives for about forty-five minutes and then fades.

Jacques Beckers and Egon Schröter obtained the first high-resolution spectra of the fibrils in 1969, at the Sacramento Peak Observatory. Their data led them to suggest that the dark fibrils are nearly horizontal and contain the Evershed flow. The bright fibrils, on the other hand, are *inclined* with respect to the dark ones. Their proposal lay dormant for almost thirty years, waiting for confirmation.

Recent observations at a spatial resolution of 200 km have vindicated them, but only at the cost of even more complexity. Alan Title and his colleagues have obtained a superb movie of the magnetic fields in a spot, at the Swedish Vacuum Solar Observatory in the Canary Islands. Their observations show that the bright and dark fibrils are indeed slightly tilted with respect to each other (see Figure 9.10) and are *inter-leaved*. The outward Evershed flow, at 0.5 to 5 km/s, is confined to the dark fibrils. Title claims the field is weaker in the bright fibrils, but Bruce Lites and his coworkers from the High Altitude Observatory claim the field strengths are the same. So even the best observations can lead to different conclusions.

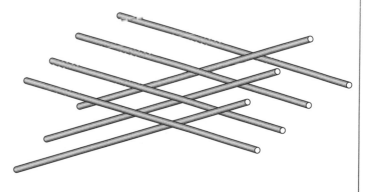

Fig. 9.10 *Interleaved dark and bright magnetic fibrils in a sunspot penumbra. (Courtesy of Kluwer Academic Publications)*

That is not the end of the complexity. Richard Muller, a very skilled French observer, has resolved the bright penumbral fibrils into chains of bright *grains*, a few hundred kilometers wide, that migrate at 0.5 km/s along a flux tube *into* the umbra. Similarly, dark grains in the dark fibrils migrate *outward* in the penumbra at about a kilometer per second. The penumbra is a virtual traffic jam!

In 1995, Thomas Rimmele at the U.S. National Solar Observatory measured the Evershed flows in the dark fibrils with extraordinary spatial resolution. His results suggest the dark fibrils are low horizontal tubes, sitting about 300 km above the photosphere, in which plasma flows up from the umbra and down beyond the edge of the penumbra.

A number of theorists have attempted to describe the Evershed flows as *siphon flows* in nearly horizontal flux tubes. If the pressure at the opposite ends of a tube differs, as is very likely, the plasma will flow toward the low-pressure end. You can think of siphon flow as suction pulling the plasma along the tube.

These latest results have to be reconciled with previous observations of migrating magnetic flux. Surrounding a spot is an annular "moat," about 15,000 km wide, in which patches of magnetic flux drift away from the spot at speeds of about a hundred meters per second. Strangely enough, the flux has both the same and the *opposite* polarity of the parent sunspot. Several ideas have been put forward to explain this odd behavior. Perhaps complete omega loops might somehow peel off the sunspot. Or possibly we are seeing some kind of wave motion along a partially submerged flux tube.

TOY SUNSPOTS

Sunspots are marvelous playgrounds for the theorist. They offer opportunities to study convection in a strong magnetic field, magnetic waves, and complex flows. And sunspots offer a stepping-stone to understand spots on stars, which are much larger, stronger, and longer-lived.

One major goal has been to fit together observations of the sunspot's magnetic field and thermal structure in a comprehensive numerical model. The model should not only predict what is seen, but also conform to some basic physical principles. Since the motions in a spot are, for the most part, slower than the local speed of sound, one can assume that hydrostatic equilibrium prevails in the plasma. Also since the plasma pressure in the spot's atmosphere is too low to confine the strong magnetic fields there, the field can be assumed to be force-free, or even "potential," that is, free of all electric currents (endnote 9.5).

Astronomers didn't tackle the tricky issue of the spot structure under the photosphere, for the good reason that they knew very little about it. But this subsurface region has attracted a lot of attention lately. As is often the case with sunspots the investigations started with a simple question: How is the umbra heated?

Everybody now agrees that the umbra is cooler than the photosphere because its strong magnetic field inhibits convection. But the umbra must receive *some* heat because although it is cool (about 4,000 kelvin), it is not stone cold. Radiation from the surrounding photosphere was shown to be totally inadequate, so the only alternative seemed to be convection in some unconventional form. How can convection work at all, however, if the umbral field below the surface is a solid rope of magnetic flux, something like the trunk of a tree?

One proposal is that hot plasma underneath the photosphere can *force* its way between the field lines of the umbra and carry heat there. In this picture, a hot tongue of plasma extends upward between field lines, dumps its heat at a level where it can be radiated away, cools, retracts to its original depth, picks up another load of heat, and repeats the process. The tongue always moves parallel to the field lines, and never turns over to sink, as a normal convection cell might. This would be oscillatory convection, and the umbral dots were thought to be evidence for it.

Further work showed, however, that the dots don't carry enough heat. So Gene Parker came up with yet another bright idea. Suppose that under the photosphere, a spot's magnetic field is not monolithic, like a tree trunk, but looks more like the dangling tentacles of an octopus, with field-free spaces between the "arms." Convective cells could then carry heat very close to the photosphere *between* the arms. This conjectural model is known as the "cluster model." It has proven very difficult either to confirm or reject, partly because it is hard to model such a nasty magnetic geometry and partly because few observations are possible for this part of a sunspot. However, as we shall see, helioseismology may give us some clues. At the present time the cluster model and the monolithic model remain competitors.

WIGGLES AND WAGGLES

Sunpots display a fascinating variety of waves and oscillations. These motions are interesting in their own right as examples of the behavior of strong magnetic fields in a dense plasma. In addition, they offer astronomers another way to probe the atmosphere of sunspots.

The photosphere of an umbra oscillates vertically, with a period around five minutes. Or more precisely, *different parts* of the same umbra oscillate independently. Since the surrounding photosphere also oscillates at five minutes (the acoustic p-waves we spoke about in chapter 5), the natural assumption is that the photosphere excites the umbra to oscillate. Although this turns out to be true, the real picture is more complicated.

As the subsurface trunk of the umbra is jostled and squeezed by the photospheric oscillations, it transforms some of the acoustic energy it receives into hydromagnetic waves. And it is quite selective about what it will accept: only those photospheric oscillations with wavelengths that fit nicely across the diameter of the umbra will do (see endnote 9.5).

In the penumbra, hydromagnetic waves travel horizontally from the boundary with the umbra to the outer edge of the penumbra, where they disappear. They move at speeds of 10 to 20 km/s, and repeat at intervals of 200 to 300 seconds. Although the waves travel primarily in the photosphere of the penumbra, they can be seen best in its chromosphere. In H alpha, a dark wave-front moves out radially from the spot and vanishes at the penumbral boundary.

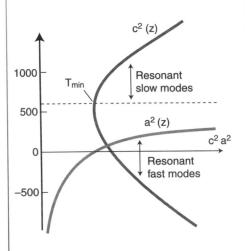

Fig. 9.11 The squares of the Alfvèn and sound speeds above a sunspot. Their crossings create two cavities in which resonant umbral waves can be trapped. (Courtesy of Kluwer Academic Publications)

Helioseismology is beginning to help us understand the subsurface structure of sunspots. In 1995 Tom Duvall, a NASA scientist, devised a clever way of estimating the amount of acoustic power a sunspot receives from the photospheric five-minute oscillations. Basically the idea is to track all the five-minute acoustic waves that enter a circular area centered on the spot and calculate their power. When the same thing is done for waves exiting the spot through the circular area, a surprising result appears: up to half of all the incident power is absorbed by the spot.

What happens to this energy isn't clear and many explanations have been suggested. For example, Henk Spruit has proposed that part of the energy the umbra receives is converted to sound waves that travel deep down the stalk of the umbra, and vanish. Just how this happens will depend on the way the subsurface part of the umbra is constructed. A monolithic umbra will behave differently than a cluster of magnetic fibrils, so it may be possible to decide which of these alternative models for the umbra is closer to the truth.

This kind of absorption of photospheric acoustic energy also occurs throughout the subsurface of an active region, wherever the magnetic fields are strong. Recently, Doug Braun and Barry LaBonte, associates of Duvall, have independently mapped the patterns of absorption in active regions, at different depths. In such an acoustic power map, the umbras show up as dark spots (i.e., regions of strong acoustic absorption) that gradually fade away at depths around 10,000 km. An active region is also embedded in a broad diffuse region of acoustic absorption. These observations are too new to have resolved any of the questions regarding the subsurface of active regions, but they almost certainly will in time.

Very recently, Braun and his colleague Charles Lindsey announced they could find sunspots on *the backside of the sun*, by analyzing sound waves that originate on that side and propagate to the front side. This is a marvelous mirror trick that promises to give us early warnings of large active regions on their way around the sun.

But it is time to move on. Active regions get their names from some of the most dramatic events the sun has to offer: solar flares. We shall consider their wonders, and their challenges to our understanding, in the next chapter.

Chapter 10

EXPLOSIONS ON THE SUN

On the morning of April 24, 1984, President Reagan was flying over the mid-Pacific on his way to a meeting in Japan. He was speaking to his aides in Washington on his high-frequency radiotelephone when suddenly, without warning, his channel went dead. When he rang up the pilot in the cockpit, he learned that all communication channels with Air Force One had suddenly shut down. For two hours the President of the United States, arguably the most powerful world leader, was incommunicado. Then, after communications had been restored, he learned that he had not been the victim of Soviet aggression, but rather one of tens of thousands who suffered the effects of a huge solar flare.

Imagine a region the size of Asia exploding with a force equivalent to billions of megaton hydrogen bombs, with plasma temperatures reaching tens of million of kelvin and a mass of a trillion kilograms hurtling skyward at hundreds of kilometers per second. In the first few seconds, a blast of deadly X-rays is beamed toward the earth. Then, a huge shock wave is launched into interplanetary space. Clouds of relativistic electrons and energetic ions follow later.

Such an outburst (Figure 10.1) with an energy content of 10^{18} kilowatt-hours, represents only a hundredth of the sun's total energy output for one second. And yet it could supply the present energy requirements of the United States for forty thousand years!

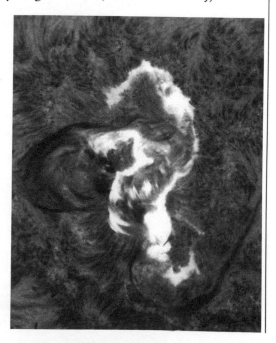

Fig. 10.1 The great eruptive flare (the "Sea Horse") of August 7, 1972. The two bright ribbons seen in H alpha lie at the feet of a magnetic arcade. At the scale of the picture our earth would appear the size of one of the small bright dots in the upper left. (Courtesy of Big Bear Solar Observatory)

As President Reagan learned, solar flares can severely affect the earth and its surrounding magnetic field. In his case the X-rays from the flare caused a sharp drop in the height of the earth's ionosphere, the layer that reflects radio signals. The charged particles a flare ejects cause harmless aurora, but if sufficiently energetic, they can destroy sensitive satellite electronics or even kill an astronaut working outside a space station.

Where does all this energy come from? How is it stored and how is it released in so short a time? Astronomers have struggled with such questions ever since Richard Carrington, the wealthy English amateur scientist we've met before, saw a flare brighter than the solar disk in 1859.

All through the nineteenth century and half of the twentieth, observers following Carrington were limited to viewing flares in visible light. Understandably, that influenced the way they thought about flares, as originating in the chromosphere and releasing only heat. However, the advent of solar radio astronomy, after World War II, led to the discovery of a variety of nonthermal flare emissions. Then starting in the late 1960s a series of rocket and satellite instruments revealed the high-temperature aspects of flares, at short X-ray wavelengths that are normally blocked by the earth's atmosphere. These new data caused a revolution in thinking about flares. It is clear now that a flare is essentially a coronal phenomenon that releases energy in several different forms, and that affects all levels of the solar atmosphere.

Flares are the offspring of active regions, especially young, growing regions in which magnetic fields are strong and changing rapidly. In fact, if one were asked, "What is a solar flare?" a short answer would be "A rapid change in a strong, complicated magnetic field."

Like active regions, flares are most numerous during the peak of the eleven-year solar cycle. A single active region can flare repeatedly, many times a day, for many days or weeks, producing flares with very different energies. As Figure 10.2 shows, weak flares are much more frequent than strong ones.

Because of their intrinsic interest and because they cause important geophysical effects, astronomers have worked hard for at least four decades to understand and to try to predict solar flares. They have pieced together a broad pic-

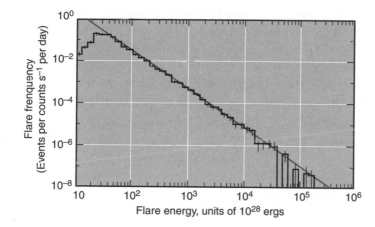

Fig. 10.2 The frequency with which flares of different energy occur. Small flares are much more numerous than large ones. The distribution is a power law. (Courtesy of Kluwer Academic Publications)

ture of the processes at work in flares, but many intractable problems remain.

PREDICTING FLARES

Because a big flare can cause a variety of problems here on earth, a small industry has grown up to predict them. Observatories around the world cooperate by sending their solar observations to central locations, like the Space Environment Laboratory in Boulder, Colorado. After processing there, the Laboratory issues a forecast for the following week. The skilled observers there have isolated a few key criteria to help them sift through the masses of data. But even with some computer assistance the process is still more art than science. It's a lot harder than predicting the weather.

Still, you don't need a degree in meteorology to guess the chances for rain tomorrow. Similarly, astronomers don't necessarily need to know all the detailed physical processes in flares to decide whether a big one is likely tomorrow. For many years, and through long experience, some observers have accumulated rules of thumb to help them forecast flares.

One of the more successful practitioners is Harold Zirin, a professor of astrophysics at Cal Tech. Zirin is a larger-than-life character, noted for his flamboyant lectures and unconventional manner. (He once used a four-letter exclamation of approval on a NOVA documentary, while watching a spectacular solar eclipse.) In the early 1960s, Zirin persuaded Cal Tech to establish a solar observatory in the middle of Big Bear Lake in California. At the time, a lake site was considered risky but Zirin's choice was proven right. Over the past forty years he and his staff have pumped out a flood of detailed analyses of solar activity. As an avid user of the observatory, Zirin has picked up his own favorite flare indicators. He publishes his estimates as "Bear Alerts" and his predictions have a good record for reliability.

How does he do it? The area of an active region, he finds, is not an infallible indicator. A large region, with several big sunspots and strong magnetic fields, can idle for days. What matters is the *complexity* of the field, how closely interwoven the positive and negative polarities are, and how fast the magnetic field is changing.

Zirin and his colleague M. A. Liggett found that some of the most energetic flares, the ones that cause the most noticeable geophysical effects, occur in regions that contain a so-called delta sunspot. It is really two sunspots of opposite polarity surrounded by a common penumbra, an extreme example of a highly complicated field. Most flaring regions don't contain delta spots, however, and one must look for other clues.

Mona Hagyard and her colleagues at the Marshal Space Flight Center in Huntsville, Alabama, have found one. For many years they have improved instruments to measure vector magnetic fields in the photospheres of active regions. If the field lines cross directly between areas of opposite photospheric magnetic polarity, the active region is unlikely to flare. But if the field lines are strongly *sheared*, so that they lie along the boundary between opposite polarities, then the region is more likely to flare. They quantified the amount of shear and learned that if a certain limit is exceeded, a flare may occur, but not always. Apparently, there are no guarantees in this business.

David Rust, a solar physicist at Johns Hopkins University, discovered another clue to when a flare may occur. Small sunspots continue to pop up as an active region grows. Rust noticed that if a new spot emerges next to another spot that has the opposite magnetic polarity, their oppositely directed fields may touch and light up a flare.

Such flare indicators are not limited to visible light. A region may heat up, fluctuate in brightness, oscillate up and down, or emit strong radio waves a few minutes before it explodes. But such activity is common in developing regions, even when they *don't* flare, so, again, there are no guarantees.

SOME FAMILIAR FACES

Although astronomers are unable to predict flares with certainty, they have recognized several distinct types. *Compact* or *confined* flares appear in tight low loops with magnetic field strengths of several hundred gauss. These flares can produce extremely hot and dense plasmas, with temperatures up to 50 million kelvin. They emit "hard" X-rays, with photon energies ranging from 20 to perhaps 200 kev, over tens of minutes.

A second type, the *impulsive* flare, often begins in a low-lying row of sheared magnetic loops. As their name suggests, they emit hard X-rays in bursts as short as a tenth of a second, simultaneously with microwaves and chromospheric spectral lines. Their maximum temperatures usually reach 20 to 30 million kelvin.

Finally there is the spectacular *two-ribbon* flare (see Figure 10.1). It typically occurs in a row of magnetic loops that lies over a cool prominence. The prominence, normally quiet, becomes violently disturbed for some reason and erupts, tearing open the loops. When the loops re-form they are extremely hot, and very beautiful (Figure 10.3).

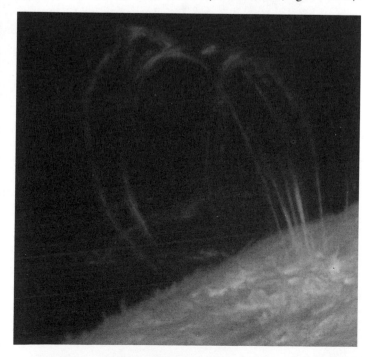

Fig. 10.3 Post-flare coronal loops, photographed in the hydrogen alpha line. (Courtesy of T. Tarbell)

They emit a full menu of X-rays, ultraviolet light, and visible spectral lines, such as H alpha. The feet of the loops also emit H alpha, thereby forming two bright ribbons on either side of the prominence. These chromospheric ribbons appear to separate sideways, away from the prominence site, as higher and higher flare loops re-form.

Such categories are useful in thinking about flares, but flares show great variety and many are hybrids that do not fit any neat criteria.

Generalities are helpful but specifics are better. Let's follow the development of a powerful and relatively simple flare, which occurred on December 16, 1991.

A CASE HISTORY

The Japanese satellite YOHKOH ("sunbeam," in Japanese) was launched in 1990 and has been observing solar activity ever since. It has been a successful collaboration between the Japanese, who built and operated the satellite, and several teams of foreign scientists, who supplied instruments to measure flare emissions over a wide range of energy. The satellite has performed as a fully equipped observatory, utilizing all its instruments simultaneously to capture the full story of each important flare. The Hard X-ray Telescope (HST) made rapid sequences of images, in X-rays with energy as high as 100 kilovolts. The Bragg Crystal Spectrometer (BCS) detected spectral lines of highly ionized atoms, such as calcium, that reveal the presence of very hot plasmas. Finally, the Soft X-ray Telescope (SXT) made dramatic flare movies in X-rays with energies of a few kilovolts.

The particular flare we describe here, which occurred on December 16, 1991, was analyzed by Len Culhane and his international team. Culhane is an astrophysicist at the Mullard Space Science Laboratory at the University College in London. He has been active in high-energy flare research since the early 1970s.

Figure 10.4 is a magnetic map of the active region before the flare, with a few calculated coronal field lines inserted. Notice how a small peninsula of positive magnetic flux intruded into the large negative area. As one might expect from the complexity rule, the flare occurred in the low-lying loops connecting these two regions.

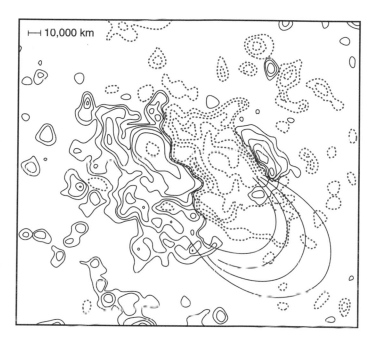

Fig. 10.4 The magnetic field in the region that produced the flare of December 16, 1991. The flare appeared near the small peninsula of positive field (solid lines) intruding into the negative portion (dotted lines). (Courtesy of Solar Physics)

Figure 10.5 shows the development of the flare in several energy bands. The whole event lasted only about five minutes. Just before the flare, the active region was heating up

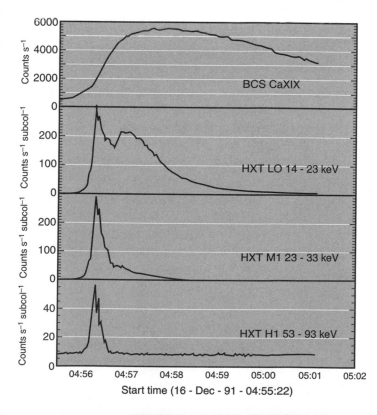

Fig. 10.5 The evolution of the flare of December 16, 1991. Hard X-rays were emitted for a much shorter time than softer X-rays. (Courtesy of Solar Physics)

to about 15 million kelvin, as shown by the gradual rise of calcium emission (Ca XIX, top panel). The flare itself started very abruptly, rising to a peak within ten seconds in an impulsive jagged burst of hard X-rays. Notice how the hardest X-rays (bottom panel) flash out in a brief spike, while the softer X-rays last longer. This is a common feature.

Fig. 10.6 The power law spectrum of hard X-rays in the December 16, 1991, flare.

The energy spectrum of the hard X-rays was a power law (Figure 10.6) with a slope of −4.1 at the peak of the burst. This means that photons with energy of 20 kilovolts were 90 times as numerous as those with energy of 60 kev. Such a power law spectrum is very different from the thermal spectrum of an isothermal plasma and requires a special explanation.

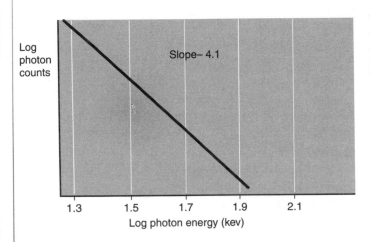

Log photon counts

Slope− 4.1

1.3   1.5   1.7   1.9   2.1

Log photon energy (kev)

A lot of the action took place at the feet of the loops. The hardest X-rays were emitted from there, and some of the hottest plasma ever recorded in a flare was produced there (50 million kelvin or three times the temperature at the center of the sun). At the peak of the X-ray burst this super-hot plasma exploded up the ends of the loops at speeds of 300 km/s and more. These are all clues to the physics of the flare, as we shall see.

Following the fireworks of the impulsive phase, a so-called "gradual phase" set in. The plasma cooled off to a mere 12 million kelvin after two minutes, but the total mass of heated plasma steadily rose, as the Ca XIX curve (top panel of Figure 10.5) shows. After five minutes the active region cooled down to its pre-flare state, apparently unharmed by its violent episode.

This flare was typical of many impulsive flares that show violent up-flows at their roots, but do not eject plasma into space. Culhane and his collaborators interpreted it using concepts that were established by the observations of two previous satellites, the Solar Maximum Mission in 1984 to 1989 and the Hinotori satellite in 1980. To follow Culhane's reasoning we need to recall these ideas.

How Flares Work: Three Stages of Development

During more than four decades of intensive research on flares, solar physicists have constructed conceptual models of flares and have begun (but only just begun) to understand their underlying physics.

The first major insight was the realization that flares are essentially magnetic phenomena. They occur only in active regions, which are characterized by strong, evolving magnetic fields. Secondly the only plausible source of flare energy lies below the photosphere, in the kinetic energy of massive convective cells. All other possible energy sources, such as gravity or heat, fail to meet the energy requirements of a large flare by a large margin.

Finally, the magnetic field of an active region can become unstable if it is stressed sufficiently by convective forces. Eventually the field collapses into a simpler, unstressed form. Its stored energy is released explosively in several forms (accelerated electrons and protons, shock waves, X-radiation) which all eventually turn into heat, and are radiated away.

The December 16, 1991, flare we described just now is typical in having three stages of development: energy storage, impulsive energy release, and then a slow thermal decay. Let's take a closer look at the physical processes in each stage.

ENERGY STORAGE

Flares are basically coronal events, but their energy derives from the motions of convective cells, deep below the photosphere. How does this energy get from deep down to high up? And how is it stored there until it explodes in a flare?

The answers lie in the way the coronal magnetic field of an active region is tied to these convection cells. Coronal loops are rooted in sunspots, in pores, and in the intergranular lanes surrounding them (Figure 10.7).

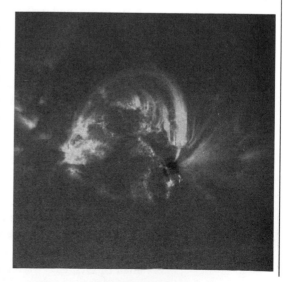

Fig. 10.7 Coronal loops in an active region are anchored in sunspots (like the dark patch), pores, and intergranular lanes. Also see color supplement. (Courtesy of the TRACE team and the Lockheed-Martin Solar and Astrophysics Lab)

Like gigantic trees, their roots extend deep below the photosphere and probably connect somehow. Convective motions at these depths are powerful enough to jostle even the strong fields of sunspots. Whenever the roots of a magnetic loop are pushed around, the loop as a whole must follow, because the field lines behave like stiff rubber tubes. Each loop in the corona therefore is constantly forced to change its shape and position. But loops are not isolated, they pack together and push against each other. So each loop is constantly crowding against its neighbors, as it readjusts to the motion of its roots.

Several theorists (among them, Eugene Parker at the University of Chicago) have suggested a number of ways in which such stressed loops might interact. Two neighboring loops could wrap their legs around each other. Three or more legs could be *braided*, like the strands of a girl's hair. A single leg might be *twisted* by the vortices between granules. Or a row of loops could be *sheared*. Although all of these are possibilities, only the sheared row has been observed so far.

Any distortion in the shape of a magnetic loop represents stored magnetic energy, which can be tapped later. As we mentioned earlier, magnetic field lines behave like stiff rubber tubing. They can be bent, stretched, twisted, and braided without breaking, but, like tubing, the more distorted they become, the stiffer they become and the more stored energy they contain. They behave this way because of the electrical currents they contain.

A magnetic field in which *no* electrical currents flow is called a *potential* field. It is "smooth" in the sense that the field lines don't wind around each other, but nest beneath each other nicely. A potential field certainly contains energy, because some agency (convection in the sun) had to work to create the field and this work is now stored in the strength of the field.

But a potential field has only the *lowest* possible energy consistent with the way its roots are arranged in the photosphere. This minimum energy is not available to power a flare.

If now convective motions work on the roots, the field lines become twisted, gain energy, and attempt to change their shape. In general this is not possible without the appearance of electric currents, that start to flow along the field lines. Once started, the currents can continue to flow almost indefinitely because the electrical resistance of the coronal plasma is as low as copper's. *These currents contain the energy that eventually powers a flare.* In fact you can think of the energy as stored either in the currents or in the twists in the field. Maxwell's equations assure us that the two descriptions are equivalent.

So to summarize, the energy to power a flare originates as the kinetic energy of convective cells, deep below the photosphere, and is stored for long periods as electrical currents in a distorted coronal field.

This whole process is rather like twisting a rubber band. In its untwisted form, the band is "potential" (we could say "relaxed"), with minimum possible elastic energy. But if you twist it the band is stressed and thereby stores the work you had to do to twist it. Moreover, the more you twist, the more the band fights back and the harder it is to twist further.

ENERGY RELEASE

As convective motions continue to stress the magnetic loops in an active region, excess energy in the form of electric currents continues to build up. At some moment, the dam bursts. Somewhere in the loop system, a particular loop is stressed beyond a critical limit. It cannot adjust to additional stress and snaps suddenly into a new configuration to relieve its strain. That tiny isolated movement bears on neighboring loops and now *they* have to adjust rapidly. A disturbance propagates rapidly, each relief of strain triggering another loop, like a row of dominoes. Within seconds, all the loops in the region are relieving their strain by reducing their twists or, equivalently, consuming their internal currents.

In 1991 Edward Lu and Robert Hamilton simulated this domino effect with a numerical model of an active region. They demonstrated that a large flare could be thought of as a collection of small ones. In effect a flare is like an avalanche, in which a small disturbance in a large unstable system causes the entire structure to collapse.

Lu and Hamilton think of flare energy being stored primarily at certain critical sites within the corona, where the magnetic field is most tangled. They represent these sites in their computer as a cubic array of sites, like rooms in an apartment building. In each second of time, the computer chooses some sites at random and sends them little packets of simulated magnetic energy. A site is only allowed to store a limited amount of energy, however. When a site reaches its prescribed storage limit, it must distribute its excess energy to its neighbors, according to some simple arbitrary rules. With this additional energy, the neighbors may also exceed their limits and have to overflow to their neighbors.

And so a simulated flare begins. The flare gathers energy as more sites dump their excess, until the rolling disturbance reaches a group of under-filled sites, which can quench the flare.

Lu and Hamilton followed this process over a long time, counting the number of flares that occurred with different energies. Small flares occurred far more often than big ones. In fact, the distribution of flare energy they computed is a power law, with the same slope as the observed one (Figure 10.2). So a simple model, with practically no physics, is able to reproduce an important characteristic of flare production. But, as the saying goes, the devil lies in the details!

Perhaps the most troublesome "detail" is the physical basis of energy release. Exactly how does a large active region dump its stored energy in just a few seconds? Solar physicists agree that the only plausible way is by *reconnection of field lines*. Figure 10.8 illustrates such an event.

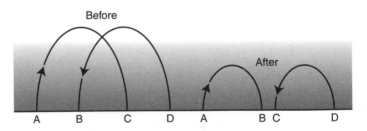

Fig. 10.8 Magnetic reconnection. On the left, two loops touch each other and reconnect to form the two separate loops on the right.

Before

After

A   B   C   D   A   B C   D

Initially, two magnetic loops with opposite field directions lie close together. The laws of electromagnetism tell us that an electric current must flow in the space between them, perpendicular to the page of the book, because the field direction reverses there. This narrow sheet of current represents excess energy that is available for a flare.

When the loops touch, their field lines change partners: they break, and instantly rejoin with new lines to form two separated loops (Figure 10.8, right). In this new arrangement, no current exists, so the excess energy must have been dissipated in the reconnection process. A natural way for this to occur is for the electrical resistance of the plasma to convert the current to heat, just as the coil in your toaster converts electric current to heat. This is one basic process in a flare, the conversion of stored currents to heat.

So far, so good. But some flares release their energy in a few minutes at most, and this short time for reconnection poses a difficult problem. In order to reconnect, two oppositely directed loops must touch. But the space between them is filled with plasma. The magnetic fields of loops could simply *diffuse* through this plasma but that would take much longer than a few seconds. As soon as a field line starts to drift through the plasma, an electric current is induced in the plasma. That current couples with the field to generate a "Lorentz" force that prevents the field from drifting further. So diffusion is slow.

In order that the field lines contact each other quickly the intervening plasma must be *removed*. The way the sun does that is to *squeeze* the plasma, and force it to flow along the field lines, like toothpaste out of a tube (see endnote 10.1). In 1964, H. E. Petschek proposes the first plausible model for such an event.

Petschek's model of reconnection has been elaborated extensively and even simulated. It's the basis of our present

Fig. 10.9 H. Petchek's geometry for reconnection of magnetic field lines. The curved (solid) magnetic lines approach from the top and bottom and reconnect in the rectangle at the center. Dotted lines and arrows show the direction of the flow. The reconnected lines are shown as vertical on each side of the rectangle and are moving rapidly away (to the left and right) from the center. Standing shocks (labeled S) separate regions of slow inflow and fast outflow of plasma.

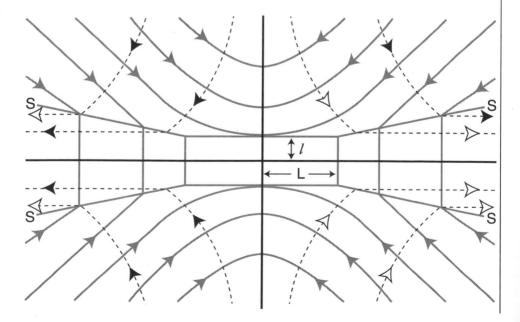

understanding of how magnetic fields reconnect and release their stored energy. But it had a serious limitation in
its original form. At a single reconnection site, the rate of
energy conversion from electric current to heat turned out
to be too small by many powers of ten to match flare observations. A Petschek flare would simply be too small and
too slow.

One obvious way out of this dilemma was to propose that
many reconnection sites combine, as in the avalanche model,
to produce a single flare. But the number of required sites
seemed improbably large. So more physics had to be introduced. The new ingredient was anomalous resistivity. Under flare conditions the electrical resistance of the plasma
can increase enormously, and this results in heating (endnote
10.2).

Reconnection, with or without anomalous resistivity, is
therefore an essential element of flare physics. Not only
does it help to explain the rapid progress of a flare, it also
offers another way to *start* one. In the Lu-Hamilton avalanche model, a flare starts spontaneously, when some critical limit of storage is violated. Another possibility is to *force*
the coronal field to become unstable. That is the idea behind David Rust's "emergent flux" scheme.

New magnetic loops are constantly bobbing up through
the photosphere in an active region. There is no reason
why a new loop has to have the same field direction as the
clusters of loops above it. When a new loop pops up with
opposite polarity and presses up against the older loops,
reconnection can occur at the interface. If the older loops
are sufficiently stressed, then this latest insult can trigger a
violent reaction, a flare.

THE SLOW THERMAL DECAY

After a few tens of seconds, the impulsive phase, with its
dramatic bursts of X-rays and explosive shocks, is over. The
overheated plasma begins to cool, and the violent motions
decay into turbulence. Heat conduction rapidly smoothes
out the wild temperature variations throughout the loop,
and the excess flare energy is radiated into space as soft X-
rays.

Spectroscopy shows that the radiation consists of the strongest lines of abundant heavy elements, such as calcium, silicon, and iron. This stage of a flare is understood fairly well,

and is not particularly controversial since it involves only thermal processes. So we'll just pass it by.

FLARE ARCHITECTURE

Since theory predicts that the reconnection zones will be too small (only tens or hundreds of meters wide) to be observed directly with existing telescopes, solar astronomers have had to guess just where such zones might appear. Guided by observations of flares, they have proposed a variety of configurations in which opposing magnetic fields might contact each other.

The simplest one is pictured in Figure 10.10, top. Here an emerging loop touches an existing loop of opposite polarity. Such situations have been observed frequently in soft-ray images from the YOHKOH satellite.

Fig. 10.10 A sample of proposed flare architectures. (Courtesy of Cambridge University Press)

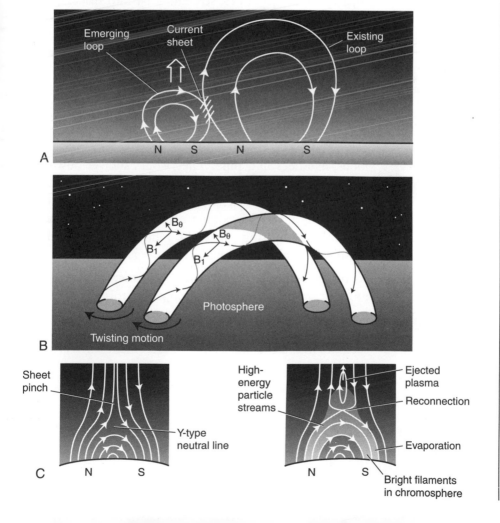

Another type of loop-loop interaction (Figure 10.10, middle) was proposed by that famous pair, Tom Gold and Fred Hoyle, in 1960. In this picture two loops have opposite longitudinal magnetic fields, but carry currents in the same direction. The electrical forces acting on the loops will push them together, allowing their oppositely twisted fields to touch and cancel.

Large two-ribbon flares display the full range of flare phenomena: the ejection of mass into space, chromospheric heating, and hard X-ray and microwave bursts. In 1968, Peter Sturrock, a plasma physicist at Stanford University, proposed a magnetic structure (Figure 10.10, bottom left) that could account for many of these observations.

He imagined that the corona above an active region looks something like a coronal streamer, with closed loops at its base and open field lines, stretching far into space, higher up. Notice that there are no twisted fields in this structure, but neither is it a potential field. An electric current sheet separates the field lines in the upper part, and this sheet can both store energy and release it rapidly when the field lines reconnect.

Sturrock proposed that some of the released magnetic energy would accelerate electrons in the plasma to high energy. These electrons would be trapped within the closed loops, like bees in a pipe. They would shower down the legs of the loops and smash into the chromosphere at the base of the loops. The plasma in two parallel ribbons would instantly light up with nonthermal hard X-rays and microwaves. Rapid heating would also follow, causing the chromospheric plasma to explode up the legs of the loops.

As reconnection proceeded, the reconnection site would move upward, and the closed loops that formed would be higher and wider. This means that the roots of newly formed loops (and their associated bright ribbons) would appear to separate in the chromosphere, in agreement with observations.

Sturrock's model has been extraordinarily successful in explaining, at least qualitatively, not only two-ribbon flares but other types, such as the one Culhane and associates analyzed. The essential idea of reconnection near the tops of loops, with the production of intense beams of electrons and protons, has been very influential. But is it realistic? As we shall see, astronomers have debated its merits for quite a while.

THERMAL OR NONTHERMAL?

A really powerful flare can emit hard X-rays, gamma rays, microwaves, and sometimes clouds of relativistic protons. How does the sun do it? How can a low-tech star produce all this high-energy radiation? Solar physicists have struggled for two decades to answer these questions. Some easy answers have proven to have unexpected consequences.

Two basically different explanations have been proposed. The hard X-rays could be emitted from super-hot plasma, just as X-rays are emitted in the center of the sun. Or alternatively, some special nonthermal process might create the X-rays. The best candidate for that process is the grazing collision of very fast (nonthermal) electrons with slow protons.

There is no doubt that the sun somehow generates nonthermal electrons and protons in flares. The electrons (with energy up to 500 kilovolts) generate nonthermal microwaves (a clear signature) and are sometimes detected in interplanetary space by satellites. Solar "cosmic rays" (protons with energy as high as a billion volts) can stream to the earth and be detected there. So, although it isn't clear how the sun generates these fast charged particles (a topic we'll return to shortly), the fact is they are there, and so they *could* produce the puzzling hard X-ray and gamma-ray photons.

A number of theorists, including the Scottish physicists John Brown and his colleague, Gordon Emslie, have explored the basic idea of electron beams. Brown teaches at the University of Glasgow. Emslie has migrated to the University of Alabama—without losing his Scottish accent, I might add.

These two Scots share a wry sense of humor and enjoy playing with ideas, especially slightly radical ones. Moreover, they like a good debate and can defend their ideas effectively. When their proposals ran into trouble they fought off the opposition for a decade, but ultimately had to modify their views. Together with Sturrock and Zirin, they have had a big influence on the way others think about flares.

In 1970, Brown developed a mathematical model to demonstrate how fast electrons can create hard X-rays in a flare. He assumed that an electron beam is accelerated somehow in the upper part of a coronal loop and streams down its legs. Unlike Sturrock, Brown thought the electrons create X-rays along the legs, not at the loop's feet. This apparently slight difference of opinion turned out to matter a great deal, and turned into a spirited debate.

According to Brown, the electrons in a beam make sharp right-angle turns around slow-moving coronal protons, and emit hard X-rays as they whiz past. If the electrons start out with a power law distribution of kinetic energy, then the hard X-rays will also have a power law distribution of photon energy, just as observed (Figure 10.2). Such a beam has another nice feature; it can also generate the microwaves that are observed. With these attractions, the model (called the "thin target model") soon drew a lot of support among astronomers.

But on further reflection, the thin target model turned out to have serious problems. Fast electrons produce hard X-rays very inefficiently. Therefore it would take a *lot* of fast electrons to generate X-rays in the amounts that are observed. By some estimates, *all* the electrons in the loop would have to be accelerated, within seconds, and would soak up the entire stock of stored flare energy. Those requirements seemed quite outlandish.

What's worse, such a powerful electron beam, moving through a coronal plasma, would induce enormous magnetic fields, which are not observed. Brown and Emslie, among others, proposed a variety of changes to the thin-target model but these had their own problems.

Gradually, opinion shifted to the "thick target" model. In this model, the electron beam is still the star player. But as Sturrock originally proposed, it produces X-rays in the dense chromosphere at the *feet* of the flaring loop, not in its legs. The beam electrons slow down as they plow into the chromosphere, and being slower, they produce X-rays much more efficiently. Therefore fewer fast electrons are needed to match the observations, and that eliminates a lot of problems. The thick target model looked more attractive.

But some astronomers had a much simpler idea. Why bother with beams at all? Suppose the hard X-rays are actually *thermal* photons, emitted from a super-hot plasma in an explosion at the *top* of a loop. The plasma would have to have a mixture of temperatures to emit a power law X-ray spectrum, and the maximum temperature would have to approach a *billion* kelvin for the hardest X-rays. (Recall that the temperature of the sun is only 15 *million* Kelvin.) But such high temperatures couldn't be ruled out without evidence, and so this model (which actually dates back to the first X-ray observations of flares) gained support too.

No model is perfect, however, and this one too has its flaws. The fast electrons in super-hot plasma can outrun the slower protons, creating, in effect, an electron beam. That raises all the objections to the beam model.

The two models, thermal vs. nonthermal, gave different predictions for the appearance of a flaring loop in hard X-rays and in microwaves. The thermal model would have both types of radiation appear at the *tops* of loops, while the nonthermal model would require the hard X-rays to appear at the loop's feet, as a beam of fast electrons collided with dense chromospheric plasma.

With the launch of the Solar Maximum Mission in February 1980, astronomers looked forward to new observations that would decide between the two models. The first hard X-ray images showed clearly that the radiation was coming from the loop feet. It appeared as though the nonthermal proponents had won the day. Later observations with the Japanese Hinotori satellite tended to confirm the role of electron beams.

But the sun is not so accommodating as to provide simple final answers. The latest YOHKOH observations, compiled by T. Kosugi, show no fewer than *six* different types of hard X-ray sources. Some flares emit hard X-rays at the tops of loops, some at the feet, some at *both* top and feet. Moreover, some flares show clear evidence for both electron beams and super-hot plasmas. Let's look at two recent examples.

A Nonthermal Flare

Len Culhane and his colleagues from the University College of London analyzed the flare of December 16, 1991, which we described above, in terms of the nonthermal model. From the spectrum of the hard X-rays they determined the total energy an electron beam would have to have. Then, using the Ca XIX spectra, they determined the energy contained in the exploding jets of hot plasma observed at the loop roots. The two estimates, 3×10^{16} and 4×10^{16} kilowatt-hours, agree quite well.

This neat result supports the idea that the primary release of energy was used to accelerate an electron beam, which then produced all the other effects during the impulsive phase. Although very hot plasma (50 million kelvin) was observed, it appeared at the feet of the flaring loops, not at the loop tops, as the thermal model proposes.

Does Culhane's analysis, and others like it, rule out the thermal model? Certainly not! It merely demonstrates *consistency* among the various observations, but does not exclude an alternative explanation, and leaves open many basic questions, such as the origin of the beam.

Other YOHKOH flares have been successfully interpreted with the thermal model. Here's an example of a flare that seems to exhibit *both* thermal and nonthermal aspects. It was analyzed by Satoshi Matsuda and his colleagues at the Japanese National Astronomical Observatory and the Institute of Astronomy, both in Tokyo.

A HYBRID FLARE

This brief "compact" flare occurred in a single loop on January 13, 1992, near the solar limb. At the peak of the flare, the entire loop lit up in soft X-rays (1–2 kev), a perfectly normal event. Somewhat harder X-rays (23–33 kev) appeared at the feet of the loop, but also at the *top* of the loop, where the temperature reached at least 20 million kelvin. And for the first time ever, the hardest X-rays (33–53 kev) were observed in a compact blob 7,000 km *above* the top of the loop (see endnote 10.3).

Matsuda's distinguished colleague, Saku Tsuneta, provided a mathematical model to clarify some of the processes that Matsuda had sketched. But one has to admit that several aspects of their explanation lack observational support. In particular they didn't observe the high-speed jets that supposedly produced the fireball at the top of the loop. Conventional models of magnetic reconnection, such as Petschek's, do *predict* such jets, but so far the evidence for them is weak. If any jets appear during a flare, they seem to flash out from the loop feet.

We see here that even the best X-ray observations obtained

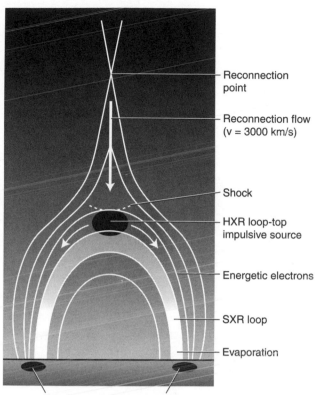

Fig. 10.11 The
magnetic geometry
proposed for the flare
of January 13, 1992.
(Courtesy of the
YOHKOH consortium)

Reconnection
point

Reconnection flow
(v = 3000 km/s)

Shock

HXR loop-top
impulsive source

Energetic electrons

SXR loop

Evaporation

HXR double footprint sources

so far fail to fully constrain a model for the acceleration of electrons, and for the heating of plasma, during the impulsive phase. A lot of imagination is still needed to explain the observations. And yet this is the best current picture we have of impulsive flares. Clearly, much more needs to be done.

PARTICLE ACCELERATION

The key problem is to understand exactly how the sun accelerates electrons and protons to huge energies within a few seconds. A satisfactory solution must explain several additional observations. First, protons and electrons are usually accelerated *simultaneously*. This suggests that the same mechanism works on both. Secondly, the energy spectrum of electrons is a clean power law, but the spectrum of the ions that produce the hardest gamma rays is messy (Figure 10.12). Finally, some flares accelerate particles in two stages, the first to tens of kilovolts, the second to hundreds or thousands.

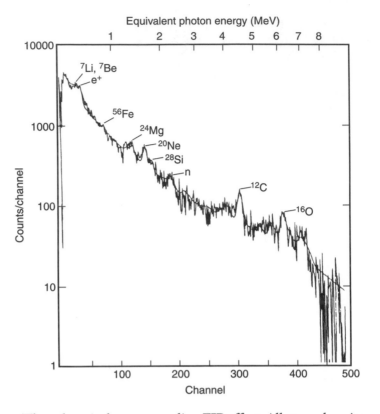

Fig. 10.12 The observed energy spectrum of gamma rays. The spikes along the curve arise from particular nuclear reactions in a flare. For example, the spike at 0.5 Mev corresponds to the annihilation of electrons and positrons. (Courtesy of Cambridge University Press)

Then there is the very puzzling FIP effect. All atoms heavier than hydrogen have more than one bound electron in their pristine state. The energy required to strip off the first electron is called the First Ionization Potential (FIP), and is measured in electron volts. (E.g., the FIP of carbon is 11.3 volts, and that of sodium is 5.1 volts.) When the ions from an eruptive flare are detected in interplanetary space, those with FIP *less* than 10 volts are found to be three to four times overabundant, relative to those ions with FIP *greater* than 10 volts. Also, in a few flares the abundance of the light isotope of helium ^{3}He is enhanced by a factor of a thousand. These effects suggest that special, charge-sensitive mechanisms exist.

For the present, theorists would be content to be able to explain the main features of particle acceleration, leaving for the future such complications as the FIP effect. They have proposed several plausible mechanisms, some of which even predict power law energy spectra.

All mechanisms ultimately depend on a strong electric field (or equivalently a voltage drop) to accelerate a charged particle. The field may be steady (at least for a few seconds), oscillating (as in some kinds of wave), or rising steadily. At the Stanford Linear Accelerator, for example, electrons are shot down a long evacuated tube by a steady electric field. Is it possible the sun works in the same way? If so, how could an electric field appear in the highly conducting plasma, which acts to short out any electric field?

As early as 1967 Hannes Alfvèn (the physicist who discovered "frozen-in" magnetic fields) and Peter Carlquist proposed their "interrupted current" model. They realized that the current in a flare loop would build up rapidly as motions of its feet stress the loop. Drifting electrons in the plasma must therefore accelerate to carry the increasing electrical current. When the drift speed of plasma electrons reaches the speed of sound at some small vulnerable site, ion-acoustic waves will develop there, causing the electrical resistance of the plasma to increase suddenly by a huge factor. As a result a large voltage drop (millions of volts) appears across the site. It is this voltage difference, and the strong electric fields it implies, that can accelerate electrons and ions to the observed energies.

Theorists have shown that once such a resistive "bottleneck" develops in a loop, it can maintain itself for at least a few seconds. But it is difficult to understand, in detail, how such a situation can arise in the first place. This model of particle acceleration has therefore fallen into disfavor.

Enrico Fermi, Nobel Prize winner and key player in the Manhattan Project to build an atomic bomb, proposed in 1949 that galactic cosmic rays could be accelerated by collisions in a turbulent, magnetized plasma. His idea has been applied to solar flares in two forms. The simpler idea ("first-order Fermi process") is that charged particles can be trapped between converging shock waves. In magnetized plasma, a shock behaves like a solid wall to a charged particle. Like a ping-pong ball bouncing between the walls of a shrinking room, a particle will pick up additional kinetic energy each time it collides with one of the shock waves. If the particle has enough energy to start with, it can reach the million-volt energy range.

A neat arrangement of colliding shocks, sufficient to energize a particle in one operation, is rather unlikely, however. More realistic is the second scheme ("second-order Fermi process") in which the particles rattle around in a turbulent field of shock waves. The particles move in random directions, gaining or losing energy at each encounter, but gaining overall. Shock waves are not hard to imagine in the violent turbulence of a flare, but whether they can trap a fast particle long enough to reach very high energy is an open question.

More exotic proposals involve interactions between some form of plasma wave and the fast particles. A plasma wave can contain oscillating electric fields. Under the right conditions a slow particle can "surf" on a plasma wave and pick up energy from it. Specialized waves with names like "Langmuir," "ion-acoustic," and "lower-hybrid" have been investigated theoretically. Both the second-order Fermi process and the lower hybrid wave can generate a power law distribution of fast electrons, a plus for each of them.

Finally, the reconnection process that lies at the heart of energy release has been invoked to accelerate charged particles. One of Maxwell's equations tells us that an electric field is generated wherever a magnetic field changes rapidly. A laboratory example is the famous betatron, a circular ring in which increasing magnetic fields induce an electric field that can accelerate electrons to hundreds of millions of volts. Magnetic fields must certainly change in reconnection but the actual geometry of the fields, to say nothing of the detailed physics, is a matter of debate. Nevertheless, some viable schemes have been proposed.

As you can see, all this is still a theorist's game. Although present observations constrain the game, they are generally insufficient either to disqualify or confirm a particular proposal. In December 2000 NASA plans to launch the High Energy Solar Spectroscopic Imager (HESSI), a space observatory instrumented to study particle acceleration and energy release in flares. HESSI will be able to observe X-rays from 3 kev to gamma rays with energy up to 20 MEV, with high time resolution. For the first time, spectroscopy and X-ray imaging will be combined, so that spectra at each point in an image can be obtained as a flare develops. Hopefully this new approach will identify the mechanisms of particle acceleration unambiguously.

A PARTING WORD

More solar astronomers work on understanding flares than on any other topic. You can see why. They present more difficult problems of interpretation than any other phenomenon on the sun. And they have important consequences for life on earth and in the space around us.

Decade by decade we draw closer to a detailed understanding of their complex physics, but some questions may never be answered to everyone's satisfaction, simply because some of the important processes are hidden in sub-telescopic regions. Predicting flares, even a day in advance, has remained an art rather than an objective science. However, research on solar flares has helped astrophysicists to grapple with flares observed on other stars. Such flares dwarf the puny objects on the sun by many powers of ten, but probably involve many of the same concepts.

Chapter 11

THE CORONA AND THE WIND

O n the evening of March 13, 1989, the lights went out all over Quebec, as cleanly as if somebody had thrown a master switch. Emergency switches melted, transformers went up in smoke, and alarm bells rang in all the control centers, indicating a major power interruption. Hydro Quebec, the main electrical utility, was deluged with thousands of calls from angry customers. It took two days to restore power everywhere.

The cause of the disaster was identified later as a huge cloud of plasma from the sun, a "coronal mass ejection" or CME. Astronomers have since learned that a large chunk of the corona tears itself loose from the sun about once a day, year in and year out. If it launches itself in an unfavorable direction, this mass of hot plasma (typically a billion tons of it) can hit the earth after a trip of three or four days. The cloud collides with the earth's magnetic field, enormous electrical currents are induced in the ground, and electric power transmission lines experience a terrific surge.

The result is widespread devastation. A CME can fuse a communication satellite's delicate electronics. A $300 million satellite was recently disabled in this way. Navigational satellites are also deranged. The damage occurs on the ground as well. CMEs have been known to destroy the remote monitors along an oil pipeline, requiring enormous replacement expenses.

Such losses have created a demand for early warnings of impending CMEs and solar astronomers are willing to help. In collaboration with their governments, they have set up monitoring systems that give at least a day's notice that a coronal cloud is on its way. Plans are afoot to launch an early-warning satellite that would orbit around a point between the earth and sun. Radio techniques to detect CMEs are also feasible, and are being discussed. And a research group has recently discovered a possible signal on the sun that an ejection is imminent.

But astronomers are dissatisfied merely to report or forecast the danger. They want to understand the underlying physics of the ejections. How and where is the required energy stored? How is the ejection triggered?

Even on a day when no mass ejection occurs, the earth is bathed in a steady flow of hot plasma from the corona. This is the solar wind. Gentler than the impulsive ejections, the wind is nevertheless a powerful influence on the earth, whistling by at speeds up to 800 km/s. It blows the earth's magnetic field into a teardrop shape and leaks back along the teardrop to cause auroras and other disturbances. Astronomers are studying how and where the solar wind escapes the corona, and are teasing apart its strange properties.

But the mass ejections and the wind are only the more dynamic features of the corona. The very fact that the corona exists at all, that it somehow maintains its temperature at millions of kelvin, has challenged astronomers since the discoveries of Grotrian and Edlén in the 1940s. Astronomers are still looking for a satisfying explanation of how the corona is heated. Recently they have taken one more step toward this goal.

At the present time, two satellites, SOHO and TRACE, are gathering new observations of the corona that bear on these three central issues. The latest images from space demonstrate that the corona is filled with extremely narrow loops and boils with activity. In this chapter we'll see how astronomers are using these new data to try to understand the corona, its dramatic ejections, and its winds. We begin with a close look at its vast architecture.

A GUIDED TOUR

Figures 11.1 and 11.2 are white light and X-ray images of the corona that reveal its main features. In Figure 11.1, an image from the LASCO coronagraphs aboard the SOHO satellite, the bright streamers extend at least to thirty-two solar radii from the sun. Other satellites have detected their influence as far out as the orbit of earth.

Between the streamers lie dimmer regions, the so-called "coronal holes." These are regions of low plasma density.

Fig. 11.1 An image of
the corona from the
Large Angle Solar
Coronagraph and
Spectrometer (LASCO)
aboard the Solar
Oscillations and
Heliospheric
Observatory (SOHO).
The small white circle
shows the size of the
solar disk. (Courtesy of
SOHO/LASCO
consortium)

Fig. 11.2 An X-ray
picture of the sun. The
dark region at the north
pole is a coronal hole
and the bright loops
are magnetic field lines
made visible by the hot
plasma they contain.
Also see color
supplement. (Courtesy
of the YOHKOH
consortium)

They also show up in X-rays (Figure 11.2) as dark patches
or lanes that cover large parts of the solar disk, especially
near the poles.

The brightest structures in X-rays are the coronal parts of
hot active regions. In high-quality X-ray pictures from the
YOHKOH and TRACE satellites, they resolve into discrete loops of hot plasma (see Figure 9.3). One of the puzzles
astronomers are trying to solve is just why only certain loops
are filled with plasma while others are empty.

Finally, you can see a spattering of so-called X-ray bright points in Figure 11.2. These are thought to cover small magnetic bipoles that are small cousins of the sunspot regions.

Everything one sees in such pictures is shaped by the coronal magnetic field, which fills all the space around the sun. Where the field is strong, as in the active regions, the field forms closed loops that trap the hot corona plasma and prevent it from expanding into space. Because the plasma is trapped, its density can build up. Dense plasma emits strong radiation, and therefore the loops appear bright in X-ray pictures.

Where the field is weak, as at the poles of the sun, the plasma is able to expand into space as the solar wind. As it accelerates outward, it combs the field lines out into long threads. Because the plasma escapes steadily along such "open" field lines, it leaves a hole in the corona, where the residual plasma density is low, and the X-ray brightness is correspondingly dim.

All the space around the sun, the "heliosphere," is bathed in a solar wind that streams outward, past Pluto, at speeds between 300 and 800 km/s. X-ray observations, made during the flight of Skylab in 1973, demonstrated for the first time that the fastest wind leaves the sun as discrete streams, like water from a hose, and that the streams originate in coronal holes. If a stream reaches the earth, it can cause a severe "geomagnetic storm" in the earth's magnetic field, and produce bright auroras.

Astronomers are working hard to understand how wind streams are related to coronal holes and how they are accelerated to their high speeds. They have learned that the largest holes, with the fastest wind streams, appear at the solar poles shortly after the maximum of the eleven-year sunspot cycle. Near solar minimum the polar streams bend down into the equatorial plane of the sun, which nearly coincides with the orbital plane of the earth. Geomagnetic storms are especially fierce at that time. Near the maximum of sunspot activity, the polar holes shrink, and smaller holes open up at low solar latitudes.

Although the fast wind has attracted a lot of attention, the slow wind, with speeds below 450 km/s, also poses problems. For instance, it still isn't clear just where the slow wind originates. It may arise from the borders of fast streams or from the tips of the solar streamers, or both.

The streamers are by far the most impressive features of the corona. Not only do they stretch far out into space, but they cover huge areas of the sun's surface. A mature streamer can extend over 150 degrees of solar longitude (i.e., 2 million km) and spread over 45 degrees of latitude.

A streamer has the shape of a flat blade or fan that straddles a magnetic neutral line on the solar surface. As Figure 11.3 illustrates, it's a hybrid between closed and open magnetic field lines. The lower part is a long tunnel of closed arches.

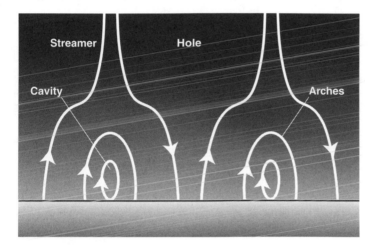

Fig. 11.3 Coronal streamers and holes. A streamer has three distinct parts: a cavity, overlying magnetic arches, and open field lines of the "stalk." Some theorists think a magnetic helix fills the cavity and is able to support the weight of a prominence.

The arches enclose a "cavity," a region of low plasma density. The open field lines in the upper part of the streamer taper down to a narrow "stalk" and then expand sideways, like a nozzle. There is good evidence that an outward flow of plasma (a slow wind) shapes this upper part of the streamer.

Fig.11.4 A quiescent prominence, photographed in the light of hydrogen alpha, shows typical fine threads of emission that probably outline the internal magnetic field lines. The prominence has a height of about 100,000 km, or about eight times the diameter of the earth. (Courtesy of National Solar Observatory)

Last in this catalogue of coronal structures are the solar prominences. They are long, narrow sheets of plasma that lie in the cavities of coronal streamers. Figure 11.4 is a photograph of one in the red light of hydrogen alpha.

Prominences really have no business residing in the corona. They are a hundred times cooler than the corona (with temperatures below 10,000 K) and a hundred times denser. This cool, dense mass appears to hang suspended in the tenuous corona and would certainly collapse if it were not supported somehow. The best guess these days is that magnetic fields in the cavity form a kind of cradle in which the prominence mass lies, as we shall see presently.

CORONAL MASS EJECTIONS

When a streamer becomes unstable and erupts, it becomes a coronal mass ejection or CME. Figure 11.5 shows a typical CME in progress, as seen in white light at the solar limb by the LASCO coronagraph aboard the SOHO satellite. In these snapshots, the shape of the CME is a distortion of the original streamer. At the outer boundary is a bright expanding shell that corresponds to the arches that overlaid the cavity. The shell encloses a dark region, the former cavity. And lagging far behind is the badly twisted solar prominence. Far ahead of the whole parade is a faint

Fig.11.5 A huge coronal mass ejection in progress on November 6, 1997, as seen by the LASCO aboard the SOHO satellite. The white circle indicates the size of the solar disk. The two white blobs on the right are the "legs" of an enormous trans-equatorial loop that erupted as a CME. Also see color supplement. (Courtesy of SOHO/LASCO consortium)

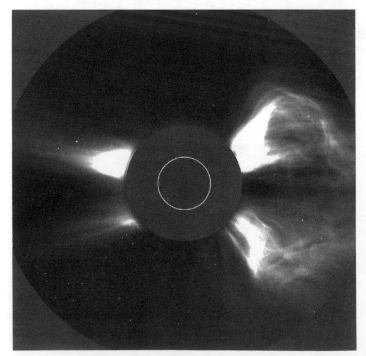

shock wave, the so-called "forerunner," that can generate a powerful radio burst at meter wavelengths.

A CME leaves the sun slowly at first, like a giant rising sleepily from bed. Although its great mass eventually reaches a speed of several hundred kilometers per second, it accelerates slowly over several hours. This behavior is quite unlike the ejections from solar flares, which lash out impulsively.

The difference lies in the root cause of each disturbance. As we saw earlier, a flare is a real explosion, in which magnetic energy is rapidly converted to heat. The heat creates an overpressure that expels jets of plasma. But not much heating occurs during a CME. Instead, the magnetic field of a streamer seems to unfold and unwind slowly, propelling the streamer's huge mass into space. The only sign of significant heating occurs long after the mass has left and the streamer is reforming. Then hot X-ray loops may form over the old site of the erupted prominence. This heating is a by-product of the main event, rather than the driver.

Although astronomers have obtained a lot of new satellite observations of CMEs, the physical cause of these dramatic events is still being debated. The principal facts calling for an explanation are the huge amount of stored energy that is slowly released and the trigger that launches the event.

Several explanations have been proposed and discarded. For example, it first seemed that a flare was needed to produce a CME. Further work revealed that if a flare is associated with a CME, it occurs long after the ejection. Indeed, most CMEs have no associated flares. The leading explanation at the moment seems to point to the cavity of the streamer as the key piece in the puzzle. To see why, we need to take a small detour in the story.

MAGNETIC CLOUDS IN SPACE

Len Burlaga at NASA's Goddard Space Flight Center has been studying the properties of interplanetary space for at least three decades, using data from spacecraft. Back in 1981, he and his colleagues obtained some exciting observations, assembled from five different satellites. In January 1978, an ejection from the sun had passed the different spacecraft at different times. At first the weak magnetic field in the flowing plasma pointed toward the north pole of the sun. Then, as the ejection passed by, the field direction gradually reversed and finally pointed south. Burlaga and associates recognized that they had detected a kind of magnetic rope floating toward the earth, in which the magnetic field is wound into a loose helix. Later observations confirmed that such "magnetic clouds" are the remnants of disrupted streamers, that is, coronal mass ejections.

Two German scientists took the next step. In 1994 they discovered that magnetic clouds that left the northern solar hemisphere produced magnetic clouds with a left-handed twist, while those from the southern hemisphere turned into right-handed twists. This relation holds up from one eleven-year cycle to the next.

At this point, David Rust enters the story. Rust is a well-known solar physicist at the Johns Hopkins University. He has a quick mind and a flair for spinning off bold ideas. He often stimulates his theoretical colleagues to flesh out his suggestions with quantitative models.

When Rust read the results of the Germans, he immediately made a connection to some remarkable discoveries about prominences by Jean-Louis Leroy and his French group, and later by Sara Martin, a scientist at the Big Bear Observatory. Martin rediscovered that the horizontal magnetic field in prominences points in opposite directions in the northern and southern hemispheres (see Figure 11.6). Rust examined Martin's observations and convinced himself that the prominence fields are helices. In fact he con-

Fig. 11.6 A map of the sun showing the magnetic fields of filaments (prominences on the disk) during one solar rotation (August 1977). The fields in the southern hemisphere generally point west, while those in the northern hemisphere point east. (Courtesy of V. Bommier)

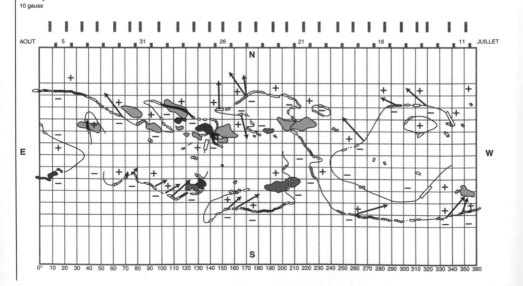

cluded that all the prominences in the northern hemisphere have left-handed helices, while all those in the southern hemisphere have right-handed helices. Rust predicted that when such prominences erupt, as part of a CME, their helices would become magnetic clouds with the observed sense of twist. He then carried out a statistical study of prominences and clouds and found the association he was looking for. It was a neat picture that could explain the origin of the helicity of magnetic clouds. But as we shall see, he may have been right for the wrong reasons.

Rust went on to sketch a speculative but highly provocative scenario for all the twisted fields on the sun. First he assembled data to show that *all* magnetic fields in the sun's atmosphere are twisted. For example, some sunspots are surrounded by spiral "whorls" which have opposite twists in the northern and southern hemispheres. Then Rust referred to published theoretical work that showed that the twists in solar magnetic fields can be rearranged but are virtually impossible to destroy. So, he concluded, the sun must cleanse itself of all the twisted fields that build up during an eleven-year solar cycle. And how does this happen? With coronal mass ejections! Rust suggested that all the twisted magnetic flux generated in eleven years floats off into space, thus ridding the sun of its unwanted helicity. A grand conception indeed, if true!

Meanwhile, Boon Chye Low at the High Altitude Observatory was also thinking deeply about coronal mass ejections. "B.C.," as he is called by friends, sparkles with enthusiasm for solar research. The son of a prominent shipbuilder in Singapore, he turned down an early opportunity to take over the family business and elected instead the life of an academic. As his mentor, Eugene Parker, discovered, he has a unique flair for constructing mathematical models of magnetic fields to interpret complicated solar phenomena. At the HAO he is having a brilliant career in his field.

Low agreed that the direction of twist in a magnetic cloud must be predetermined in a streamer before it erupts as a CME. But unlike Rust, he placed his bet on the faint cavity. Low guessed that the low-density cavity contains a helical magnetic "rope" that extends along the full length of a streamer (something like a "Slinky" toy). In fact, according to some of his hydromagnetic models, the massive prominence within a streamer *hangs* at the bottoms of the coils of the magnetic helix, like a sheet on a clothesline. Low suggested that the helix in the cavity possesses opposite twists in opposite hemispheres and turns into a magnetic cloud when the streamer erupts. He has shown how the imprint of the cavity's helical coils on the chromosphere could give Rust the impression that the prominence is a helix. Low also provided a physical cause of a CME (see endnote 11.1).

In Low's scenario, the energy to propel the CME is contained in the twist of the magnetic helix, not as the heat of an explosion. His scheme also accounts for the slow acceleration of the CME. Low agrees with Rust's idea that CMEs carry off most of the magnetic flux and helicity that the sun creates during its eleven-year activity cycle, but he extends the picture to include the reversal of polarity of the solar poles.

Low's model has many appealing features. Where it runs into trouble is in the origin of the cavity helix. Low maintains that the helix is formed *below* the photosphere and rises into the corona, fully formed and twisted in the proper sense. The problem is understanding how such a helix could maintain its form as it rises through the turbulent convection zone. But once the helix exists in the corona, the rest of Low's scenario proceeds quite naturally. He and Sarah Gibson have recently simulated the whole event with a three-dimensional, time-dependent model.

Active regions are also sources of CMEs. They also contain prominences under a streamer-like structure. In X-ray images from YOHKOH, such as Figure 1.2, the region's coronal magnetic fields are outlined in hot plasma. The fields often bend into an "S" shape, a so-called sigmoid that connects the opposite ends of the prominence. Several astronomers, among them David Rust, have noticed that the shape of the S is reversed in the northern and southern hemispheres. This is another example of a handedness, or "chirality" that depends on which hemisphere you consider.

In March 1999, Richard Canfield and colleagues announced, with some fanfare, a potentially useful warning sign that a CME will erupt from an active region. They examined many X-ray images from YOHKOH and learned that the S-shaped fields *brighten* a few days before a CME occurs. The alarm is not fool-proof ; not every CME has one, but every brightening does precede a CME.

The brightening may be interpreted as the effect of an increasing electrical current that distorts and strengthens the original field. When the magnetic pressure in the region builds up sufficiently, the twisted field erupts.

To confirm or reject these theoretical ideas, astronomers would like to have independent measurements of the strength and shape of coronal magnetic fields, particularly in the crucial streamer cavity. The observational problem is that the cavity is so dim that there isn't enough magnetic signal to measure. Even worse, the X-ray corona is so transparent in general that it is virtually impossible to disentangle the foreground corona from the background. The situation is somewhat better in active regions near the disk's center, where the X-ray structures seem to give a pretty good representation of the 3-D field. At least it has been possible to match such observations convincingly with magnetic models of the region.

HEATING THE SOLAR CORONA

Since the 1940s, when Grotrian and Edlén demonstrated that the corona has temperatures exceeding a million kelvin, solar physicists have struggled to come up with a convincing mechanism for heating the corona. We mentioned briefly (in chapter 6) that the first good idea—heating by sound waves that are generated in the convection zone—failed. Such sound waves will turn into shocks in the chromosphere, and heat it, but won't survive to heat the corona.

So other types of waves, magnetic waves, were investigated. One type, the magnetosonic wave, was soon discarded. These waves are attractive because they compress the plasma and form shocks that could heat the corona. Unfortunately, they are also strongly refracted in the diverging coronal magnetic field. A train of such waves rising into the corona would simply bend down again and dump their energy in the chromosphere.

Another promising wave type is the Alfvèn wave. Joe Hollweg, an astrophysicist at the University of New Hampshire, has been one of their strongest proponents. As you may remember, an Alfvèn wave is a traveling twist or bend in a magnetic flux tube. It carries nonthermal energy, which could heat the corona. Hollweg was impressed by the discovery in 1971 of weak Alfvèn waves in the solar wind. Hollweg suggested that these waves were the remnants of much stronger waves that heat the low corona.

There are several problems with this suggestion. First, no definite proof has been found that Alfvèn waves exist in the low corona. Even high-resolution X-ray images of isolated coronal loops from SOHO (Figure 11.7) do not show the oscillatory brightness or velocity variations (the "smoking gun") that one would like to see if Alfvèn waves were present.

Fig. 11.7 Coronal loops resolved in a spectral line at 17.1 nm of Fe IX and Fe X, corresponding to a temperature of 1 million kelvin. Also see color supplement. (Courtesy of the TRACE team and the Lockheed-Martin Solar and Astrophysics Lab)

However, this result may only reflect a pervasive uncertainty in the periods of Alfvèn waves. The conventional wisdom decrees the waves would be generated by granulation and have periods similar to granule lifetimes of about ten minutes. Certainly no such waves have found. But the solar wind's waves have periods of hours and so there might exist a broad spectrum of periods, extending down to minutes or even seconds. In 1990, Jay Pasachoff, an astronomer at Williams College, looked for very short period waves during a total eclipse of the sun and claimed he found some.

Pasachoff holds the world record for observations of total solar eclipses. At last count he had brought experiments to at least thirty eclipses, in all parts of the world. After the eclipse of 1980, he reported brightness fluctuations of the green coronal line (530.3 nm) with typical periods of a few seconds. Some theorists, including Eugene Parker, complained that such periods were unlikely from a physical point of view. But Pasachoff has continued to improve his equipment and has tried to detect short-period waves at other solar eclipses. So far he still hasn't obtained convincing results. Other astronomers also continue to search.

There are other ways to try to identify Alfvèn waves but the existence of these waves is still not definitely proven (see endnote 11.3).

So Eugene Parker, always ready with an idea, offered another suggestion. Instead of waves, he said, consider tiny coronal flares. We have described previously (chapter 10, on flares) how the slow twisting and braiding of magnetic flux tubes by photospheric convection induces electric currents to flow along the magnetic field. Eventually these currents can reach a critical strength and dissipate their energy by heating the local plasma, as the current in your toaster heats the coils.

This process is a candidate explanation for flares of all sizes, from the largest (with total energy of 10^{18} kilowatt-hours) to the smallest "microflares" (with energy of 10^{13} kilowatt-hours). Parker simply proposed to extend this same heating mechanism to the corona. Suppose, he said, that the slow twisting of coronal fields by granules releases heat in thousands of "nano-flares" with energies as low as 10^{11} kW-hr.

Why not? To test this idea, Toji Shimizu looked for such tiny flares in the YOHKOH X-ray data. In a careful study he found some as small as 10^{12} kilowatt-hours but not nearly enough of them. Hugh Hudson, a bright, cheery solar astronomer working with the Japanese YOHKOH team, had pointed out in 1990 that, to heat the corona, the number of flares smaller than 10^{12} kW-hr must be hundreds of times larger than those observed at 10^{12} kW-hr. Shimizu found there simply aren't enough of them to do the job. (For the latest report, see endnote 11.3.)

At this stalemate, Jack Scudder, an expert plasma physicist at NASA's Goddard Space Flight Center, launched a radical proposal. Scudder stated boldly that the energy the corona emits is not deposited in the corona, so there is no point in looking for it there. Instead, he proposes that some nonthermal process heats a small fraction of *chromospheric* electrons and ions to speeds that correspond to coronal temperatures. These particles, and only these, have enough energy to rise, against the force of gravity, to coronal heights in the solar atmosphere. (Just as a ball must be thrown with sufficient speed to reach a third-story window.) In effect gravity filters out the slowest chromospheric particles. So when one samples the coronal plasma, one finds, not surprisingly, that it consists of energetic charged particles.

Scudder argues that several kinds of nonthermal processes in the chromosphere can produce an overabundance of very fast ions and electrons, relative to the number a Maxwellian distribution would contain. If he adopts an artificial distribution of particle speeds in the chromosphere, he can predict the filtering effect of gravity and then the nonthermal velocity distribution in the corona.

Two of Scudder's critics have tested his idea by calculating the spectra his model corona would emit. They find that his model can account for the observed intensities of spectral lines formed below a temperature of a 100,000 kelvin. That is a real achievement that would solve a long-standing problem in solar physics. But unfortunately, Scudder's model fails to reproduce the intensities of *coronal* lines, formed at 1 million kelvin and more. So Scudder's explanation for the high temperature of the corona fails in a very basic way.

That leaves us, at the moment, in a quandary. No explanation of coronal heating seems to fit all the requirements. However, some small steps have been taken recently.

LIGHT AT THE END OF THE TUNNEL

Eric Priest is a professor of applied mathematics at St. Andrews, the oldest university in Scotland. It is located near the hallowed St. Andrews golf course, which attracts devotees from around the world. Once a week, undergraduates used to parade around the walls of their city in their bright red gowns—until recently, when the walls became unsafe. Priest teaches these young students as well as guiding the research of a covey of bright young astrophysicists.

In 1998, Priest and his colleagues analyzed the temperature distribution in an isolated coronal loop that was observed in X-rays by the YOHKOH satellite. They found the distribution is best modeled by a heating source that is distributed uniformly along the loop. This result, they claim, rules out several proposed heating mechanisms.

In particular they dismiss any mechanism, such as reconnection of local magnetic fields, that injects heat at the feet of a loop. Trapped waves of any kind are apt to deposit heat nonuniformly, so they too are dismissed. Turbulent dissipation of many current sheets, proposed as early as 1975 by William H. Tucker, and more recently by Parker, is a possible surviving mechanism, in their view.

Other groups are also working hard on this heating problem and one at least has come up with a conclusion that contradicts that of Priest and company.

The SOHO satellite is equipped with a magnetometer and two telescopes that produce images in selected spectral lines. In 1997, a team from Stanford University and the Lockheed-Martin Solar and Astrophysics Lab combined observations from these instruments to make two connected discoveries.

First, the magnetometer revealed that thousands of small magnetic bipoles are constantly popping up at the solar surface. These small, low loops form what the researchers call a "magnetic carpet" (Figure 11.8). As new loops emerge, they collide with existing ones. The team claims to see reconnections among the constantly jostling loops, with the

Fig.11.8 Small magnetic loops extend into the corona from magnetic bipoles in the photosphere, forming the "magnetic carpet" in the quiet sun. The bright patches were imaged in the 19.5 nm line of Fe XII by the EIT on SOHO, and show where the low corona is heated. Also see color supplement. (Courtesy of SOHO/ MDI/EIT/CDS consortia)

release of stored magnetic energy.

Secondly, images from SOHO's Extreme ultraviolet Imaging Telescope (EIT) and the Coronal Diagnostic Spectrometer (CDS) show vigorous heating and motions of the plasma at the feet of coronal loops. The team concludes they have found the long-awaited evidence of coronal heating by small-scale reconnections. Notice that the heating would take place at the feet of coronal loops, not uniformly along the loop as Priest and associates concluded.

Finally, TRACE observations of active regions also indicate that heating occurs at relatively low heights (10,000 to 20,000 km) in very thin loops. The heating fluctuates in minutes, and wanders through the corona as the loops' feet meander. Slow heating by electric currents may be responsible. So, although these discoveries can be counted as real progress, there is still room for discussion!

The Solar Wind

In 1953 Ludwig Biermann, the German astrophysicist we've met before, became interested in comet tails. The tails generally pointed directly away from the sun, regardless of the direction the comet nucleus took in its orbit. Occasionally blobs of dust and plasma in the tails could be seen to accelerate along the tail. Biermann realized that since the pressure of sunlight is constant, some other force must be present. He proposed therefore that the sun occasionally emits streams of "corpuscular radiation" (or what we would call electrically charged particles) that collide with the molecules in the comet tail and push them along the tail. This was the first indication that the sun emits a wind of charged particles.

Eugene Parker took the next important step in 1958. He reacted to a model of the solar corona published by Sydney Chapman, a well-known geophysicist. Chapman had assumed the hot corona was static and computed profiles of plasma temperature and density along a radius from the sun. His predicted density near earth agreed with Biermann's estimates from comet tails.

The problem with Chapman's model was that it predicted a finite coronal pressure at very large distances from the sun. Parker pointed out that the known interplanetary gas pressure was far too small to match Chapman's predicted pressure, and so the corona could not be static—it must expand into space.

Parker started over again with the new assumption that the corona must expand, and built several models with the requirement that the pressure at infinite distance from the sun must vanish. The basic cause of the expansion, he showed, is the thermal energy in the corona. A hot corona wants to expand into empty space and is only restrained by gravity. But the strength of gravity falls off at increasing distances from the sun. If the corona can inflate to a height where the gravitational pull of the sun is sufficiently small, the plasma can then flow freely into space.

Figure 11.9 shows one of Parker's simple isothermal models along with the profile of wind speed it predicts. At great distances from the sun, the flow would reach speeds of 500 km/s or more.

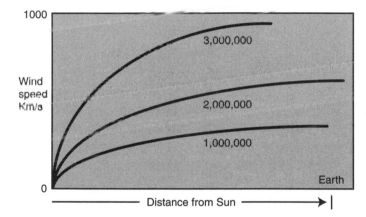

Fig. 11.9 A sketch of Parker's isothermal wind models, showing the increase of wind speed with distance from the sun. (After E. N. Parker)

Parker's prediction remained just that and nothing more until 1962, when the Mariner II Mission to Venus returned direct measurements of the interplanetary medium. Parker's prediction was completely confirmed and the solar wind became one of the principal topics of study for astronomers and geophysicists.

Table 11.1 summarizes some of the chief properties of the wind. We have already mentioned that the wind speed varies in space and time, from as low as 300 to as high as 800 km/s, and that the fast wind originates in the coronal holes.

In order to reach a wind speed of 800 km/s, a Parker model would require a coronal hole temperature of about 3 million kelvin. When new observations of coronal holes became available during the flight of Skylab, astronomers received a rude shock. The holes were *cooler* than the stream-

Table 11.1 **The solar wind at the Earth's distance from the Sun**

	LOW SPEED	HIGH SPEED
Proton density (cm^{-3})	12	4
Bulk velocity (km s^{-1})	327	702
Proton temperature (K)	34,000	230,000
Electron temperature (K)	130,000	100,000
Energy content (%)		
Gravitational	74	40
Kinetic	21	53
Thermal	1	1

ers by as much as a million kelvin. A typical hole temperature is about 1.4 million kelvin, much smaller than the 3 million a Parker model needs to explain the high wind speed. What was even more puzzling was that the flux of kinetic energy carried in a wind stream was at least two or three times larger than any wind previously known. How could a fast stream achieve its ultimate speed?

The answer seems to be some form of extended heating or acceleration in a wind stream. Parker had assumed for simplicity that all the energy in the wind is deposited at some low height in the corona and that heat conduction spreads this energy outward to maintain the flow. The new results on wind streams demand some additional mechanism to continue to heat (or push) the wind, far from the sun.

All the usual mechanisms to heat active regions were reexamined to see whether any could apply to the fast wind streams. Parker's nanoflare mechanism would not work because it requires closed field lines to store magnetic energy, whereas the field in the wind streams is open. Only Alfvèn waves offered any hope.

In 1988, George Withbroe, a solar astronomer at the Harvard-Smithsonian Center for Astrophysics, gathered all the existing observations of the fast solar wind and calculated a number of wind models that match them. He found that at least 20% of the energy input to a fast wind stream must be deposited beyond the distance from the sun (around 2 solar radii) at which the wind reaches the local speed of sound, about 200 km/s. The remaining 80% must be injected below that distance.

Parker realized that Alfvèn waves, generated somehow in the photosphere, would not dissipate until they had traveled far along the open field lines of a wind stream. In this sense Alfvèn waves are ideal for the purpose of extended heating of the stream. But the observations of spectral line broadening show that, if Alfvèn waves exist at all in a coronal hole, they contain barely enough energy (i.e., the necessary 20%) to heat the distant parts of a wind stream and not enough energy to heat the inner parts.

Parker concluded that a source of heating other than Alfvèn waves must account for the 80% of the total energy input that is deposited low in the coronal hole. In an inspired guess, he proposed that this energy is released from the reconnection of small-scale fields at the borders of supergranulation cells. You may remember that small magnetic bipoles emerge constantly in the centers of the cells and are swept to the borders. There, the opposite polarity magnetic fields drift, merge, split, and reconnect in a furious flurry of activity. Conceivably, Parker suggested, enough energy is released in this frantic activity to heat the lower parts of coronal holes.

This was pure conjecture on Parker's part. But in February 1999, new evidence that supports his idea was reported (see endnote 11.4).

This is by no means the end of the story but only the beginning. We need to confirm that the "magnetic carpet" scenario really does heat the corona and does provide the energy for the solar wind. But recent progress has been very encouraging.

Chapter 12

"Like a Tea Tray in the Sky"

Some of the prettiest objects on the sun, to my eye, are the delicate, lacy clouds in the corona, called prominences. In Figure 11.5 we saw one at the limb of the sun, photographed in the red light of the hydrogen H alpha spectral line. This cloud could hang there, slowly changing its shape over several hours, like a veil in a gentle breeze.

If we look straight down on the surface of the sun we see prominences as dark, narrow "filaments" (Figure 12.1). These are two names for the same object. These two views, at the limb and at the sun's center, imply that a prominence (or filament) is a thin sheet, a curtain of gas in the corona.

The wonder is that such an object can exist for any length of time, because it is like an ice cube in an oven. Since it

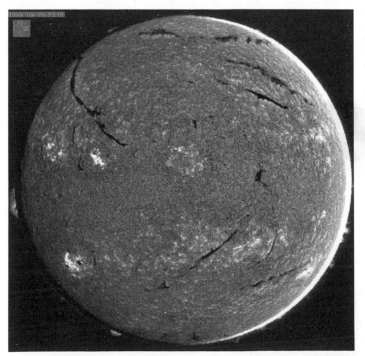

Fig.12.1 Quiescent filaments seen in absorption against the solar disk, in the H alpha line. The tail of the long inclined one in the lower right is growing into a channel of aligned fibrils that mark the magnetic polarity inversion line. (Courtesy of Big Bear Solar Observatory)

emits the characteristic visible spectrum of hydrogen atoms, we know its temperature must lie around 10,000 kelvin, while all around it, the corona rages at 2 million kelvin. Common sense suggests the prominence should evaporate instantly, but it doesn't.

By laboratory standards, a prominence is nearly a complete vacuum, with a plasma density of less than 10^{11} atoms per cubic centimeter. But despite its lacy appearance, a prominence is a hundred times as dense as the corona around it. So it is less like a cloud than a brick wall, hanging in midair for many days. Why doesn't it fall out of the corona? The accepted view is that the prominence mass lies in some sort of magnetic hammock or cradle that is strong enough to support its weight. But as we shall see, the shape of the magnetic field and its connections to the corona are still very controversial.

In fact, although prominences have been studied off and on since the middle of the nineteenth century, and although all kinds of theoretical models have been proposed, many of their most basic properties remain unexplained to everyone's satisfaction.

For example, a large prominence contains a billion tons of hydrogen gas, about a tenth of all the mass in the quiet corona. Where does all this mass come from? What kind of magnetic shape is required to support the mass? How does this shape form and why does it fill with cool prominence gas? And once formed, how does a prominence get the energy it needs to survive? Even at a mere 10,000 kelvin, it would radiate away all its heat in an hour if it had no steady source of energy.

Most prominences don't just fade away, they erupt violently at some point in their life cycle, sometimes as part of a coronal mass ejection. But a new prominence will often form at the same location of the erupted prominence. Astronomers want to know why and how these events occur.

In this chapter we'll see what the experts have to say about these intriguing objects.

SOME BASICS

Like all objects in the corona, prominences are creatures of the sun's magnetic field. They invariably form along the line on the photosphere that separates regions of opposite magnetic polarity, the so-called "polarity inversion line" or PIL. In Figure 12.1 all the dark filaments lie along such inversion lines.

Several distinct filament types can be seen in this H alpha image. At the extreme top is a so-called polar crown. These loose, tufted objects form at very high solar latitudes just after the maximum of the sunspot cycle. Most of the dark filaments in the rest of the picture are "quiescent" filaments that grow in the spaces between decayed active regions. Notice the long inclined one at 10 o'clock. Its tail is growing into a well-defined "channel" of aligned chromospheric fibrils that follow the polarity inversion line. Then, at 9 o'clock, just left of the bright region, you can see a couple of very thin, "active region" filaments.

Such filaments can grow in length either by filling an empty channel with cool mass, or by linking up with existing filaments. As we shall see, two filaments must have the same horizontal magnetic field direction in order to link up in this way. Some quiescent filaments grow to enormous lengths, winding over large portions of the sun's surface.

The longest prominences lie between the legs of coronal streamers. We can see an example in Figure 12.2, a photograph of the February 16, 1980, eclipse. The streamer at 2 o'clock is seen end-on. At its base, a so-called cavity (a dim

Fig. 12.2 The 1988 eclipse, showing a long streamer at two o'clock, nearly end-on. A bright prominence lies within the dark tunnel-like "cavity" at the base of the streamer. (Courtesy of the High Altitude Observatory)

region of reduced plasma density) surrounds a low, very bright prominence.

At one time, astronomers thought prominences formed by collecting mass from the surrounding corona, evacuating a cavity as a result. But it turns out that the volume of the cavity could never have contained sufficient mass to form the prominence we see within it. So astronomers are forced to look for other ways in which a prominence can acquire its mass.

A prominence seen in H alpha seems to be a thin veil, made up of many thin threads and knots, with what look like empty spaces between them. Actually the spaces contain plasma that is too hot to emit H alpha. Figure 12.3, obtained with the Coronal Diagnostic Spectrometer aboard

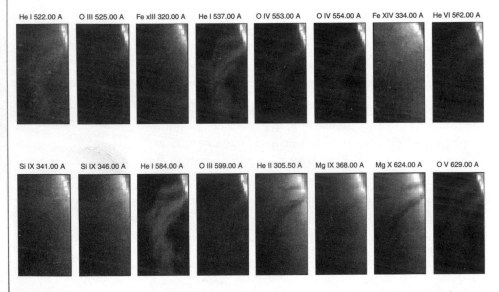

Fig. 12.3 A single prominence as imaged in different spectral lines. Each line corresponds to a different plasma temperature (e.g., 20,000 K for He I 584 A; 200,000 K for O V 629 A; 2,000,000 K for Fe XIII 320 A). Also see color supplement. (Courtesy of SOHO/ CDS consortium)

the SOHO satellite, shows a prominence in several spectral lines that were chosen to display different temperatures. Remarkably, its gross appearance doesn't change much as we look in different lines.

In Helium I, it resembles an ordinary H alpha prominence at 10,000 K. It looks much the same in the two Oxygen IV lines (at 100,000 K) and also in the Oxygen V line (at 200,000 K). In Mg X 62.4 nm (1,000,000 K) the cool dense prominence is seen as an *absorbing* cloud against the bright corona. And finally in the true coronal line Fe XIII 32.0 nm (2,000,000 K) there is no sign of the prominence.

Astronomers are trying to understand how plasmas at different temperatures are arranged in a prominence. Evidently the hot and cool plasmas must lie so close to each other as to appear almost co-spatial in images like Figure 12.3.

Two possibilities have been discussed at length. In the "core-sheath" model, a prominence contains nearly identical cylindrical threads of plasma. The plasma temperature in the core would be low (say, 6,000 kelvin) and would rise to perhaps 2,000,000 in a hot, thin sheath that surrounds the core. In the "many-thread" model, each coarse structure would consist of several threads with different temperatures.

Deciding between these alternatives is not so easy. Individual threads are often at the limit of spatial resolution, possibly a few hundred kilometers at most, so that mapping a sheath is not feasible. And because a prominence is semitransparent, one sees many different threads piled up, one behind the other, making it difficult to decide whether they have distinct temperatures.

Of course it's possible both models are correct to some extent!

INVENTING A SKY HOOK

Once astronomers realized how dense a prominence is, compared to the corona, they had to explain why it doesn't immediately fall to the surface. Two German scientists at the Max Planck Institute of Astrophysics, Rudolf Kippenhahn and Arnulf Schlüter, offered a solution in 1956 that remains accepted to this day. Their idea stems from the properties of plasma in a horizontal magnetic field.

The plasma in a prominence is a good conductor of electricity, about as good as sea water. When a blob of prominence plasma starts to fall through a horizontal magnetic field, an electric current is induced in the blob. (This is the principle of the alternator in your car, except copper wires, not plasma, are involved there.) The induced current interacts with the field to produce a so-called Lorentz force, an electromagnetic force that resists the falling motion. Voila! The blob is suspended, or at worst drifts down at a few millimeters per second. Another way of picturing this process is to think of the field lines as horizontal elastic strings, whose tension supports the prominence mass, like washing on a tight clothesline.

In 1956, practically nothing was known about the magnetic

field in and around a prominence. So although the Kippenhahn–Schlüter model of prominence support was plausible, it couldn't be accepted until it had passed an observational test. In the early 1970s, Jean-Louis Leroy and his colleagues at the Meudon Observatory, near Paris, set out to provide the necessary data.

Leroy's specialty is measuring the polarization of solar spectra, a technique that requires incredible patience, skill, and attention to the tiniest of details. He will spend hours, if necessary, meticulously combing out the bugs in an experimental setup to insure success. As a result, he has earned a reputation for making measurements at the limit of possibility. In the early 1970s, Leroy turned his talents toward observing the so-called Hanle effect in prominences. It is an indirect way to measure prominence magnetic fields. Endnote 12.1 describes the way it works.

So by measuring the decrease of polarized light, due to the Hanle effect, Leroy was able to determine both the strength and the direction of the internal magnetic fields in prominences. He observed literally hundreds of quiescent prominences, during two solar cycles, using different spectral lines. At the same time, his coworkers gradually improved the quantum-theoretical description of the Hanle effect.

To interpret their observations they assumed, as do all theorists today, that the prominence plasma must lie in a shallow dip in a nearly horizontal field. Then they could determine that the horizontal field vector makes a small angle (20–40 degrees) with the long axis of the prominence. In addition, they determined that the field strength ranges up to about 30 gauss, or about twenty times the maximum strength of the earth's field. As we shall see shortly, these results have stimulated a variety of theoretical models of prominence magnetic fields. But perhaps the most far-reaching of Leroy's results concerns the global pattern of prominence fields.

THE GLOBAL PICTURE

Leroy's observations showed that the magnetic field in a typical quiescent prominence lies nearly along the long axis. It's as though the prominence sheet has a stiff "spine." David Rust, the innovative astronomer we've encountered before, first pointed out that the directions of the axial fields in filaments (the disk counterparts of prominences) are organized over the surface of the sun in a clear, predictable fashion.

Figure 11.8 illustrated this organization. In the northern solar hemisphere the axial fields of high-latitude filaments all point in one direction, toward the east. In the southern hemisphere the direction is reversed, toward the west. Notice also that filaments at mid-latitudes have the opposite directions. Rust, relying on his sharp physical intuition, predicted that in the next solar cycle, beginning in 1985, all these directions would reverse. The northern high-latitude filaments would point toward the west, for example, and the southerners toward the east.

When Leroy measured more prominences in the new cycle, Rust's conjecture proved to be exactly correct. (Another example of the power of insight and intuition!) That raised the question of exactly how the sun manages to arrange its filaments in this systematic way.

At this point, Sara Martin enters the story. She has an incredible eye and memory for the details of solar magnetic structure. For many years she worked at the Big Bear Solar Observatory as a skilled analyst and observer. Then in 1997 she decided to break out on her own as a free-lance scientist. Working with talented undergraduates of Cal Tech, Martin has uncovered several basic clues to the global organization of filaments.

To begin with, Martin developed a method of inferring the direction of the axial magnetic field in filaments, not from the Hanle effect, but solely from the pattern of dark H alpha fibrils in the filament channels. Following a tip from Peter Foukal, a former post-doc at Big Bear Observatory, she noticed that the fibrils pointed in opposite directions on the two sides of a filament. When she compared these fibrils with detailed magnetic maps she learned that each fibril terminates in a patch of the "wrong" polarity for that side of the filament. (You see, most of the magnetic flux on one side of a filament is positive, most on the other is negative.) The fibrils perversely attach to *negative* patches on the positive side and *positive* patches on the negative side.

She then made an inspired leap of imagination. She identified each fibril as the visible, dark end of a magnetic field line that splits off from the body of the filament and terminates in the photosphere. The magnetic polarity at the visible end of a fibril then determines the direction of the field on that line. All the field lines identified in this way have a single consistent direction along the axis of the filament. So from a comparison of a fibril map and a photospheric magnetic map she was able to find the direction of the filament's axial field.

Next she noticed a striking pattern of the axial fields. Take a look at Figure 11.8 again. Pick out a particular filament in the northern hemisphere and imagine that you are standing on the photosphere with your feet in the region of positive magnetic polarity. Notice that the axial magnetic field of your filament is directed toward your *right*. That applies to almost all of the northern filaments, both at high and at mid-latitudes. If you repeat the exercise for the filaments in the southern hemisphere, you will find the axial fields point to your *left*, with few exceptions.

Martin was the first to notice this regularity. She named right-pointing filaments *dextral*, and left-pointing filaments *sinistral*. She showed that nearly all filaments in the northern hemisphere are dextral in this sense, and nearly all those in the south are sinistral. What's more, this property persists unchanged from one solar cycle to the next, although, as Rust pointed out earlier, the *absolute* directions (east or west) do alternate.

A SEARCH FOR CAUSES

What could cause this remarkable set of regularities? The issue is far from settled but two contending groups are searching for answers. We could name them the "Subs" and the "Supers." Like Rust, the Subs think the regularities arise from some combination of differential rotation and magnetic reconnections deep *below* the photosphere. The Supers on the other hand hold out the possibility that the same or similar processes occur *above* the photosphere, while a filament forms.

Aad van Ballegooijen, a mild-mannered Dutchman, has belonged to both groups at different times. He is an extremely clever theorist, working now at the Harvard-Smithsonian Center for Astrophysics in Cambridge, Massachusetts. In his first incarnation as a Sub, he collaborated with Eric Priest and Duncan MacKay, theorists at the Scottish University of Saint Andrews, to invent the scheme shown in Figure 12.4 for a high-latitude filament in the northern hemisphere (see endnote 12. 2).

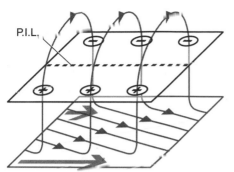

A Preparation (dextral)

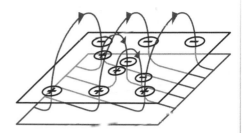

B Filament channel formation

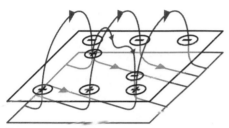

C Filament formation

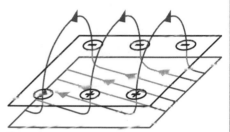

D Preparation (dextral)

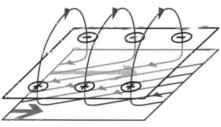

E Preparation (sinistral)

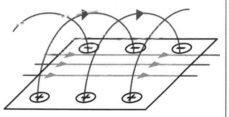

F Dextral field before eruption

This scheme is a plausible sequence of events that could explain the origin of dextral and sinistral filaments, but it has the disadvantage that the key processes below the photosphere are unverifiable.

Van Ballegooijen, like many skeptical scientists, has an aversion to such "black boxes," whose operations are invisible. So eventually he joined the ranks of the Supers, and tried to find a scheme that would produce the observed filament regularities with an observable, or potentially observable, machinery.

Working with Eric Priest and a grad student, he constructed a sophisticated computer simulation of the dispersion of magnetic flux, from active to quiet regions, over the whole sun, and over several years of the solar cycle. The simulation includes three essential processes: differential rotation,

Fig. 12.4 The formation of a dextral filament, according to the conceptual model of E. Priest et al. (Courtesy of E. Priest)

meridional flows toward the sun's poles, and reconnection of coronal magnetic field lines that are stressed by the photospheric flows.

Figure 12.5 shows their results. As the cycle progresses, the polarity inversion lines of several old active regions become linked and then distorted into what van Ballegooijen calls "switchbacks" and the coronal magnetic field becomes quite complicated. In Figure 12.5 we see the direction of the coronal field in a northern switchback, projected onto the solar surface. Notice that a filament forming on the bottom part of the switchback (on the thin dashed line)

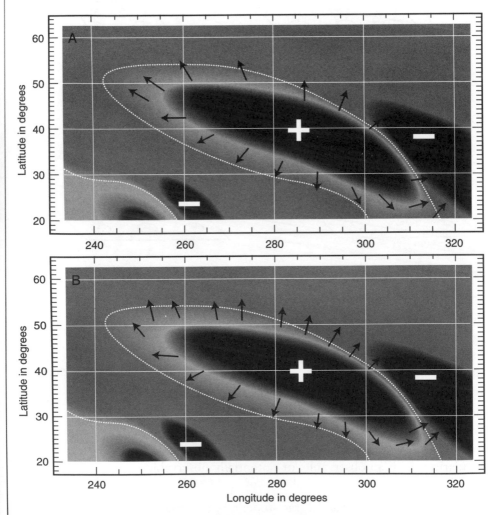

Fig. 12.5 As the solar cycle progresses, photospheric inversion lines link up to form "switchbacks." This figure shows one example in the northern hemisphere (the thin dashed line) from a simulation of the solar cycle. The arrows show the direction of the coronal magnetic field. A dextral filament would form on the lower part of the switchback, but an unrealistic sinistral filament would form on the upper part. (Courtesy of A. Van Ballegooijen and N. Cartlege)

would have the correct dextral character, but a filament on the top part would have the incorrect sinistral character. In short, the simulation fails to completely reproduce the behavior of the sun.

Van Ballegooijen has reluctantly concluded that the ordinary, well-known plasma flows, at and above the photosphere, are insufficient to explain the global pattern of filaments. He has since rejoined the ranks of the Subs and is trying to reformulate their proposals.

AGAIN, THE DEVIL'S IN THE DETAILS

Leroy's results have launched a spate of theoretical models of prominence magnetic fields. Figure 12.6 shows a few of these. All of them support the heavy prominence mass in shallow concave field lines, either across or along the length of the prominence sheet. They differ principally in the details of how the prominence field connects to the corona and to the photosphere. Some assume the large-scale field is an array of arches, some assume a twisted magnetic rope. At present, many schemes fit the observations, but since coronal magnetic fields are too weak to measure directly, no single model has been confirmed or rejected.

As you can see in Figure 12.6, all the models treat a prominence or filament as a uniform sheet, with little attention to its structural details. In contrast, Gillaume Aulanier and his colleagues at the Paris Observatory have been able to reproduce a particularly important part of real filaments.

If you glance at Figure 12.7, you will see that this typical quiescent filament has "barbs" or "feet" that seem to extend sideways from the main sheet of the filament, down toward the chromosphere. Sara Martin and her Cal Tech students noticed yet another global regularity in the arrangement of these barbs. It can best be described with an analogy.

Imagine that a filament is a super highway and that the barbs are the off-ramps. If the filament lies in the northern solar hemisphere, all the off-ramps will appear on the right-hand side of the driver, regardless of which direction along the highway the driver takes. As you may have guessed by now, the directions reverse for a filament in the southern hemisphere. So all dextral filaments have right-bearing barbs, and similarly all sinistral filaments have left-bearing barbs. Another puzzling global regularity!

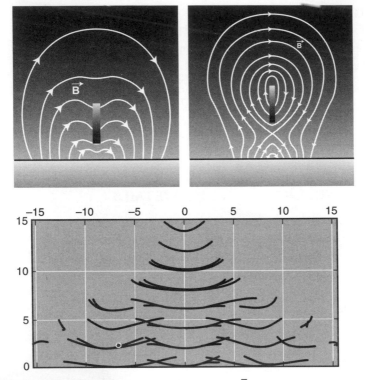

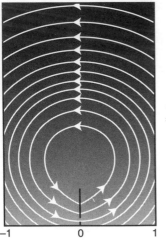

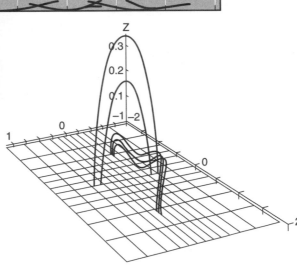

Fig. 12.6 A sampler of prominence models. Reading from upper left, clockwise they are: Kippenhahn-Schlüter, Kuperus-Raduu, Aulanier et al., Antiochus, Klimchuk and Dahlberg, and Low-Hundhausen.

Martin and her friends found yet another important clue. Most of the magnetic flux on each side of a filament has a single polarity, positive on one side, negative on the other. But a barb has the "wrong" magnetic polarity, positive on the negative side of the filament, negative on the positive side.

Aulanier and friends used this discovery to generate numerical models of filaments with realistic barbs. They arranged small patches of "wrong" polarity on each side of a PIL. That insured that some filament field lines will touch the photosphere at the sides of the main PIL and will create barbs. Figure 12.6c shows an end view of one of their models. As usual the filament mass is supported in shallow "dips" in the field lines. The barbs are the wing-like extensions on each side of the main filament sheet.

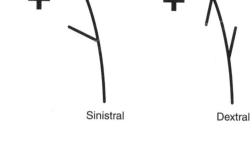

Sinistral   Dextral

Fig. 12.7 Filaments on the disk, showing barbs extending from the horizontal spines of quiescent filaments. Dextral and sinistral filaments have distinctly different barb patterns relative to their positive polarity sides.

Aulanier's models actually resemble real filaments, but leave open the question of why the patches of wrong polarity are located so as to produce the observed "off ramp" effect. Aulanier chose these locations in his models in order to get the right answer.

THREADS, KNOTS, AND DRIFTS

All prominence models, Aulanier's included, are static. The next big task in modeling prominences is to incorporate the observed motions of the plasma. Because plasma is constrained to move along field lines, not across them, a blob of moving plasma traces a real field line. By observing plasma motions, an observer can hope to test some of the models of the magnetic field. In some ways the observations already contradict the best models.

Some of the earliest (and among the best) H alpha movies of prominences were made in the 1950s by Richard Dunn at the Sacramento Peak Observatory. Figure 11.8 is a frame from one of his films. Notice the fine threads, knots, and blobs of mass that comprise a real prominence—a far cry from the uniform sheet of theoretical models. In Dunn's movies, discrete knots of emission, less than 1,000 km in size, seem to drift up or down along well-defined tracks, at speeds of a few kilometers per second. Spectroscopic observations of the Doppler effect show that some of this motion is real and not apparent.

Oddbjorn Engvold, a Norwegian astronomer who is an expert on prominences, studied Dunn's films intensively. He concluded that blobs of prominence plasma were sliding along well-defined *vertical* magnetic field lines. How can such a field be reconciled with the classical picture of a static prominence suspended in *horizontal* field lines?

Several ideas come to mind. The motions one sees in prominence movies could be wave-like, and not involve a net flow of mass. Like a tree in the wind, a prominence could merely be rippling. That possibility is pretty well ruled out by the movies.

The downward slide of discrete blobs might be explained if they could slip through the magnetic field lines that support them. Then gravity would take over and accelerate them downward. But although some known plasma effects would allow a blob to drift through a magnetic field, the predicted speeds would be far too small to agree with the observations.

How about jets of prominence plasma shooting up and falling down along vertical field lines? The problem there is that one sees rather steady motions, without the accelerations gravity would impose on such jets.

Perhaps the horizontal field lines themselves are drifting up and down, carrying their loads of prominence mass? But in that case one might expect the field lines to tangle, with nasty consequences for the stability of the prominence.

Engvold and his colleagues in Oslo are looking into yet another possibility. The discrete knots in a prominence might be prevented from falling out of the corona by the pressure of Alfvèn waves along vertical field lines. A slight change in the pressure could cause a knot to rise or fall slowly. This idea is highly conjectural at this stage and flaunts the established picture of horizontal field lines everywhere, but it is a faint hope.

BACK TO THE ICE CUBE

We've left dangling the question of how a cool prominence manages to survive in the inferno of the corona. Allied to this question is the problem of where the prominence obtains its supply of energy.

The first question is more easily answered. A cool prominence in contact with hot coronal plasma could easily heat up and merge with the corona within a few hours, if it weren't for its skeleton of magnetic field lines. Heat pouring in from the corona is the prominence's greatest danger, but the field strongly inhibits heat conduction.

Heat is carried primarily by fast coronal electrons. When

such an electron enters the cool prominence sheet it is forced to spin around the prominence's magnetic field lines. Eventually it loses its thermal energy by collisions with the local prominence electrons. The bulk of the prominence plasma would be shielded from such electrons, and stay cool, but a hot temperature transition zone might develop like a skin on the prominence sheet.

However, the corona also bathes the prominence in X-rays, and these are also a source of heating. If the prominence were a uniform sheet of cool plasma, the X-rays would be absorbed in a thin sheath at its surface. But we know the prominence is more like a fish net, with wide open spaces for the radiation to penetrate deep into the prominence. Astronomers are still debating just how serious X-ray heating could be, and whether in fact it could supply the prominence with its necessary supply of energy.

Other sources of energy are possible, of course. Among these are Alfvén waves and DC electric currents, both generated in the photosphere by the action of plasma motions on prominence magnetic fields. No final answer is available yet.

BIRTHS AND DEATHS

Solar astronomers share a wry joke: "Everything really interesting on the sun either happens during the night or on the back side of the sun." The birth of filaments is a good example of this unhappy situation. Only rarely can one be caught in the act of formation and even then, the physics involved is not very clear.

However, Victor Gaizauskas, a Canadian astronomer, got lucky. He directed the Ottawa River Solar Observatory until it shut down in 1985 for lack of funds. He still has access to the thirty-year archives of the Observatory and continues to mine them for new science. One day in 1995 he stumbled upon a sequence of H alpha images that show the formation of a filament channel and dark filament, from start to finish.

The filament channel formed during two days, between a rapidly growing active region and an old decayed region (see Figure 12.8). A pattern of dark curved fibrils, something like the pattern of iron filings around a bar magnet, extended from the east side of the active region. These fibrils would get darker for a while, then disappear, only to form

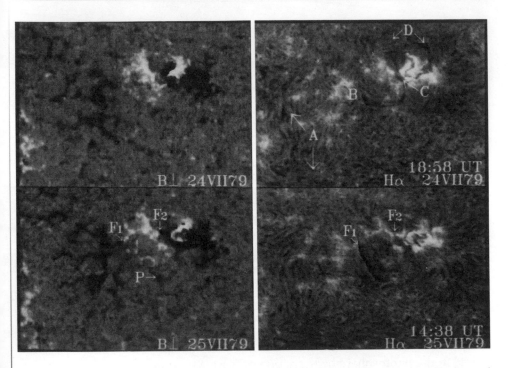

Fig. 12.8 Formation of a filament, 24-25 July, 1979. Left column, magnetic fields, right column, H alpha. (Courtesy of V. Gaizauskas)

again within hours. Gradually a broad zone of aligned curved fibrils organized itself into a stable channel. Then, *overnight*, a small, dark filament appeared, fully formed, in the channel.

Duncan MacKay, a young Scottish student of Eric Priest, collaborated with Gaizauskas to construct a computer model for the formation of the channel. The model shows how force-free magnetic fields emerging from the active region press against the adjacent fields of the decayed region, forcing some curved field lines to lie flat on the chromosphere (Figure 12.9). These field lines have the same shapes as the fibrils observed in the filament channel and presumably are their sources.

At a critical point, a new patch of magnetic field lines emerged within the channel. Evidently, this patch acted as a magnetic anchor for one end of the dark filament that formed overnight. MacKay's model does not attempt to describe how dark absorbing plasma filled in the empty field lines of the channel but does provide reasons why the filament formed when and where it did. Basically, the nearly horizontal field lines in the channel provide the magnetic support for the filament's weight.

This study goes a long way toward an understanding of how filaments form between active regions. More work is

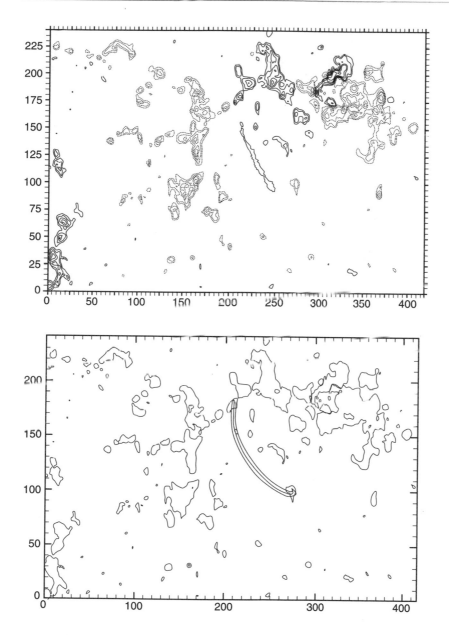

needed to explore how the high-latitude filaments are born, far from active regions.

So much for the birth of filaments. How about their deaths? Unlike old soldiers, most filaments do not simply fade away, they erupt! We have described earlier how a large filament within a helmet streamer becomes buoyant and strains to leap into space as a part of a coronal mass ejection. The filament is held down for a while by the streamer's curved field lines, like a tethered balloon. But eventually, the teth-

Fig. 12.9 Top: Photospheric magnetic fields near a filament (solid lines=positive). Bottom: MacKay's model for the filament. (Courtesy of D. MacKay)

ers weaken and the filament is free to escape. It seems that even very small, transient filaments behave in this way, although much less is known about their superstructures.

Once a filament erupts as part of a mass ejection, a new filament can appear in the same place within a few hours. Although the magnetic field in the corona is torn apart by the ejection, the *sources* of the field in the photosphere remain unaffected. Given time, the coronal field will respond to these sources and build up again to much the same shape it had before the ejection. Then, under the "proper" circumstances, the coronal field will fill again with plasma (presumably by injection from the photosphere) and create another filament. So a filament may be reborn, time and time again.

Chapter 13

SOLAR AND STELLAR CYCLES

T he sun waxes and wanes, like a flowering plant, in a cycle of about eleven years. If we could take a photograph of the sun every month for a decade and then play them back as a movie, we'd see the remarkable changes that occur so gradually. In the sun's period of growth, the corona would blossom into a halo of streamers, active regions would crowd to the surface, filaments would parade across the disk, and flares would sparkle almost continuously. Then gradually all this frantic activity would subside. The solar disk would become as blank as a dinner plate and huge holes would open in the corona.

Sunspots were the first, and in some ways still the best, indicators of the sun's activity cycle. If you will recall, Heinrich Schwabe discovered the cycle by patiently observing the numbers of sunspots for a full decade. Only later did astronomers discover that practically all the features we can see in the sun's atmosphere, and indeed some in its interior, change in step with the eleven-year cycle.

There is a marvelous order to the whole affair that astronomers are eager to understand. Figure 13.1 may help you recall the behavior of sunspots, summarized in Hale's rules. It shows that spots often emerge in bipolar pairs, roughly along an east-west line, a "leader" that is further west than

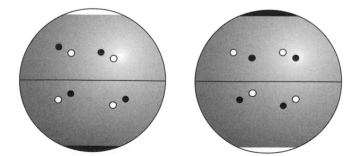

Fig.13.1 Hale's empirical rules of sunspot behavior. In an odd-number cycle, all leaders (followers) in the northern (southern) hemisphere have positive (negative) polarity and the north (south) pole has positive (negative) polarity. In the following even cycle, all these relations reverse.

the "follower." All the leaders in the northern solar hemisphere have the same magnetic polarity during a cycle, and all those in the southern hemisphere have the opposite polarity. These polarities, as well as the polarities at the solar poles, reverse during the next cycle.

There are practical reasons for studying the cycle. Many geophysical effects are driven by the sun and therefore follow the sunspot cycle. Some of them can cause expensive damage to delicate equipment or disrupt vital services. For example, charged particles from flares can fry satellite electronics and fast wind streams can whip up geomagnetic storms that blot out radio communication. So a demand is growing for better understanding of the cycle. At a minimum, the telecommunication industry would like to have a reliable forecast of the start, duration, and severity of the next cycle.

The sun is not alone in cycling. As we shall see, some stars like the sun also have stable cycles, while others have none. Why? What factors determine cyclic behavior in a star? Do stars switch off their cycles occasionally? (The sun may have done so in the past, as we'll see in the next chapter.) These are the kinds of questions astronomers would like to answer, whether or not they have practical applications.

In this chapter we'll explore the underlying physics of the solar cycle and examine some clues to stellar cycles.

CRYSTAL BALLS

Figure 7.3, the record of sunspot numbers, has teased solar astronomers for decades. It's not strictly periodic, like the swing of a pendulum, but it definitely isn't random either. It seems to be a mixture of periods, some as long as eighty years, perhaps longer. Could one learn anything about the causes of the cycle by analyzing the past record?

Many people have tried. They have decomposed the patterns of Figure 7.3 into a family of sine waves, in order to extract the different periods and amplitudes that are present in the record. Then they tested their results by predicting the next cycle's characteristics.

The results are fairly disappointing—a straightforward extrapolation of the past doesn't work too well. Either there is a random element in the cycle, or the future depends on the past in a very subtle way that is not yet revealed in the record. In fact the record is pitifully short, a mere 250 years of continuous observations, and that is probably insufficient to reveal all the subtleties.

Some helpful clues have been extracted from the record, however. For example, a cycle that rises very steeply is likely to reach a high maximum. The rate of rise and the maximum are correlated. And a long cycle is apt to follow a shorter one, as though the sun tries to maintain its average production of sunspots over several decades. But such clues offer no guarantees for a reliable prediction.

No, there is no free lunch. In order to predict the behavior of the sun's activity, one needs to understand the underlying physics. Several solar physicists have had the courage and tenacity to try to build numerical models of the sun, incorporating their best ideas of the physics involved, in order to gain a deeper understanding of how our star really works.

Their story does not have a happy ending just yet. We begin with a father and son team of astronomers.

BABCOCK'S PICTURE

Horace Babcock and his father Harold invented the modern magnetograph, a device for measuring and displaying the weak fields of the sun. Over several decades, as Horace used his machine to study the sun, he thought deeply about the sunspot cycle. In 1961 he offered a sketch of how the cycle might work, based on his knowledge of Hale's rules.

Figure 13.2 summarizes his picture. The cycle starts with a weak magnetic field that lies in the convection zone, and extends from pole to pole—a so-called *poloidal* field. Differential rotation in the convection zone wraps the north-south field lines around the sun's equator (Figure 13.2, Stage 2), creating a band of *toroidal* magnetic ropes in each hemisphere. With sustained wrapping, the field strength in a rope gradually increases, reaching some critical limit (about 3,000 gauss) in a few years.

As Ludwig Biermann first suggested, ropes are buoyant because the plasma density inside them is much lower than outside. Babcock suggested that for a rope to rise to the surface, it must first reach a critical field strength. Then, magnetic loops pop up buoyantly at random places along a rope and rise out of the convection zone and into the at-

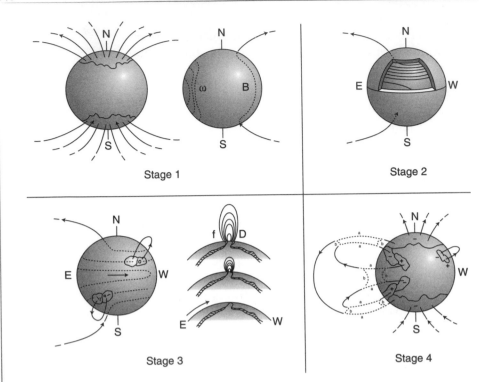

Stage 1

Stage 2

Stage 3

Stage 4

Fig.13.2 Horace Babcock's conceptual scheme for the origin of Hale's rules during the sunspot cycle. Stage 1 begins with a global polar field. In Stage 2, differential rotation winds the polar field like a spool of thread, creating a toroidal field. Omega loops rise to the surface to form bipolar spot groups in Stage 3. These rise into the high corona, reconnect, and leave the sun. The polar field remains. (Courtesy of H. Babcock)

mosphere (Figure 13.2, Stage 3).

Notice how the field direction along the ropes points east in the north and west in the south. Therefore, where a loop cuts the photosphere, we see two sunspots of opposite polarity and the leaders in the northern hemisphere have the opposite polarity to those in the southern hemisphere. In this way, Hale's laws are automatically satisfied. Moreover, because the ropes at high latitude were wrapped first, they reach the critical field strength before those at lower latitude. That would explain the appearance of new sunspots at lower and lower latitudes during the cycle, creating the "butterfly diagram." As endnote 13.1 explains, the follower spots cause the magnetic poles of the sun to reverse polarity.

What happens to all the new magnetic flux generated during a cycle? Babcock's idea, based on observations, is that the flux from each leader spot merges with and cancels the opposite polarity flux of the follower spot just west of it. At the end of the cycle, the leaders from the north and south emerge close to the equator, where their fields annihilate each other.

That, in crude outline, is Babcock's scheme. It has many appealing features that are consistent with observations, but it does have flaws. Perhaps the most serious is the way in which a new poloidal field is created. In Babcock's scenario this field connects the sun's poles in the corona and must return somehow to the convection zone. That is physically implausible, because the flux is buoyant and resists submergence.

Gene Parker, whom we have met frequently along our journey, proposed the following alternative, as early as 1955, and it has been accepted as closer to the truth.

How to Make a Poloidal Field: The "Alpha Effect"

As a buoyant convective cell rises toward the photosphere, it expands horizontally. The well-known Coriolis force (endnote 13.2) deflects its horizontal motions, so that the cell begins to turn in a slow spiral, like the draining water in your bathtub. The spiral has a clockwise twist in the northern hemisphere and counterclockwise twist in the southern hemisphere. The same effect creates cyclonic storms in the earth's atmosphere.

Now imagine that the cell carries a magnetic loop toward the surface. Then the loop will also turn in a slow spiral and develop a component in the north-south direction (see Figure 13.3). In other words, it gains a *polar* component.

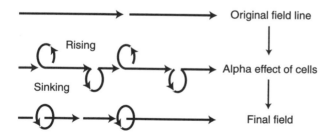

Original field line

Alpha effect of cells

Final field

Fig. 13.3 The alpha effect in the northern hemisphere. An original toroidal field line is pointed east-west. As an omega loop in the field line rises, the Coriolis effect twists the loop clockwise, creating a poloidal field from a toroidal one. The twist is reversed in the southern hemisphere. (Courtesy of Kluwer Academic Publishers)

Parker suggested that many loops twisted in this way could reconnect their polar fields to re-create the global polar field that existed at the start of the cycle. Notice that all this occurs in the convection zone. There is no need for a coronal poloidal field to submerge, as in Babcock's scheme.

DYNAMO WAVES

Parker went on to develop a simplified mathematical theory of how the sun's magnetic cycle works. Basically the cycle is an oscillation between a global *poloidal* field and a global *toroidal* field. As we have seen, differential rotation creates toroidal ropes from a poloidal field by wrapping the field lines around the sun. This is called the "omega effect." At the same time, the Coriolis force, acting on rising magnetic loops, creates a poloidal field, by the so-called "alpha effect." As the loops rise buoyantly to the surface to create sunspots, Hale's laws of sunspot polarity are automatically satisfied.

Parker's theory also explains the butterfly effect, in which new spots appear at lower and lower latitudes as the cycle advances. This is the result of a *migrating dynamo wave*, as illustrated in Figure 13.4. In endnote 13.3 I give a crude geometric description (following Peter Hoyng) of how a

Fig.13.4 Dynamo waves carrying magnetic ropes toward the equator during the sunspot cycle. See text. (Courtesy of Kluwer Academic Publishers)

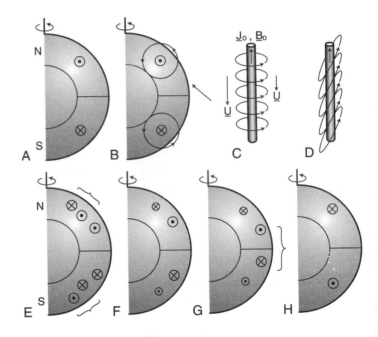

toroidal field is modified by the alpha effect and by annihilation of opposite polarities. These effects combine to force the migration of the toroidal field (and the sunspots it creates) toward the equator, the "butterfly" diagram.

All these physical principles can be put into mathematical form and a computer can then crank out a solar cycle. Figure 13.5 illustrates the output of such a simulation. The computer had to be tuned to give the correct eleven-year period, but otherwise it gave a fair representation of sunspot migration and the reversal of the field at the poles.

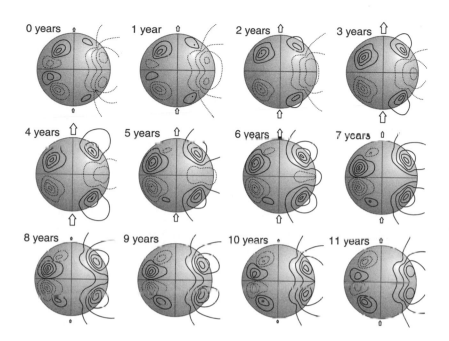

This simulation of the cycle is *kinematic*, meaning that the motions it employs were assumed, not calculated from a deeper theory of convection or differential rotation. But the results were encouraging.

BETTER MODELS?

When really powerful computers became available, a few solar physicists were able to tackle a *dynamic* model of the solar cycle. It would include the physics missing in a kinematic model, that is, a description of convection and differential rotation based on first principles. Peter Gilman and his associates at the National Center for Atmospheric Research, in Boulder, Colorado, were the first to attempt this awesome task on Cray computers (endnote 13.4).

Gilman's 1983 model was a great success. It predicted the latitude variation of rotation speed that is observed at the photosphere. But it also predicted that the bottom of the convection zone rotates slower than the top.

That last prediction would cause Gilman a lot of grief, because kinematic dynamos, such as the one shown in Figure 13.5, obtained the desired results only if the bottom of the convection zone rotates *faster* than the top. Sure enough, when Gilman used his rotation results to model dynamo waves, he found the waves migrated toward the *poles*, not the equator.

Fig. 13.5 An example of a kinematic model of the sunspot cycle. In each frame, the left half shows toroidal fields, the right half shows poloidal fields. Solid lines are pointing out of the figure (left side) or clockwise poloidal field (right side). Notice the realistic approach of toroidal field (with sunspots) to the equator as the cycle advances. (Courtesy of Kluwer Academic Publishers)

Gary Glatzmeier, at the Los Alamos National Laboratory, thought he could improve on Gilman's models. He had even more powerful computers at his disposal and could do a better job of describing the compressibility of the plasma and the turbulence. So he spent several years developing and testing his computer code. When he was finished in 1985, he confirmed Gilman in all essential respects: with the theory they had adopted, sunspots would migrate toward the poles not the equator.

Needless to say, these failures were very distressing. The best minds had used the best physics and the most powerful machines and had come up with the wrong result. Where was the flaw in their thinking?

In the midst of all the post-mortems, the first helioseismic maps of solar rotation were published. As we saw in chapter 5, these maps showed that at each solar latitude, the top and bottom of the convection zone have about the *same* rotation speed. In other words, there is no *radial* variation of angular speed, contrary to the assumptions of both kinematic and dynamic modelers!

Other problems with solar cycle models began to surface. They centered on those magical magnetic ropes that turn into sunspots. As we know, the field strength in a large spot can reach 3,000 gauss. In order for the field to build up to this level the sun would need sufficient time to wrap a weak polar field many times around the equator. But Gene Parker showed in 1975 that a buoyant rope would rise through the convection zone in much less than eleven years and would reach the surface with fields much weaker than 3,000 gauss. How could the rope remain submerged long enough?

Even worse, a rope with a field strength of only 3,000 gauss would be chewed to bits by the vigorous convective cells it passes on its way to the surface. A field strength of at least 10,000 and possibly 30,000 gauss would be needed to survive the trip.

But that raises yet another problem. Such strong fields would be stiff. They would resist being twisted by cyclonic convection. The alpha effect would be suppressed and the dynamo would sputter to a halt.

There seemed to be too many conflicting requirements for the dynamo to work. By the early 1990s, the whole subject of the solar cycle was in turmoil. What to do?

A New Look

Several theorists had pointed out that the ideal place to store a magnetic rope was just *beneath* the convection zone, at the boundary with the radiative zone. Convective motions would "overshoot" the boundary slightly, push down on a buoyant rope, and keep it submerged almost indefinitely. The field there could reach 100,000 gauss without bobbing up.

Another nice feature of this overshoot region is its *shear*. If you flip back to Figure 5.12, you can see that the radiative zone rotates as a rigid body, while the rotation speed of the convection zone varies in latitude. This means that at some latitudes, the convection zone rotates faster or slower than the radiative zone and that implies the zones *rub against each other* near their common boundary. This shearing motion is ideal for stretching magnetic field lines and for amplifying their strength. Here is a possible way to generate fields as strong as 100,000 gauss.

So the overshoot region has the potential for solving two problems associated with magnetic ropes: how to generate their strong fields and at the same time prevent them from rising prematurely. In addition, the overshoot region has shearing motions that helioseismology tells us are absent in the convection zone! All signs began to point to the bottom of the convection zone as a new frontier.

Enter Gene Parker again. In 1993, Parker constructed a toy model of the interface between the convection and radiative zones, to see whether a solar dynamo could work there. Recall that the dynamo depends on two effects: *stretching* poloidal field lines in order to generate strong fields (the omega effect), and *twisting* the resulting toroidal field to re-create poloidal fields (the alpha effect). In all previous models, both effects were located in the *same* place, the convection zone. Parker's idea was to separate them, placing the omega effect below the interface and the alpha effect above it.

Below the interface, in the radiative zone, there are no convective motions and therefore little or no turbulence. Therefore diffusion of magnetic field through the plasma occurs *very* slowly. That allows the azimuthal shearing motions to amplify the field strength in a rope as high as 100,000 gauss.

Above the interface, and in a narrow layer near it, turbulent diffusion is vigorous. It helps the field below the interface to *leak* slowly into the convection zone. The field that escapes in this way is considerably weaker, perhaps only 10,000 gauss. With such an intermediate field strength, a rope can resist being shredded and still be twisted by cyclonic convection.

Parker's trial dynamo showed a way out of the dilemmas that had accumulated with the old style of model. But it was extremely simplified. Recently, two theorists at the High Altitude Observatory, Keith MacGregor and Pierre Charbonneau, have incorporated Parker's ideas in more realistic models. They adopted outright the patterns of differential rotation in and below the convection zone that have been revealed by helioseismology. Unlike Parker's model, the rotation near the interface in their models change smoothly from the convection zone to the radiative zone.

Fig. 13.6 Results from an interface dynamo model. In the curved convection zone the meaning of the lines is the same as in Figure 13.5. (Courtesy of K. MacGregor, Ap. J 1997)

Figure 13.6 shows the results of one of their models. It is similar to Figure 13.5 in that toroidal ropes migrate to the equator and the polar fields reverse, as observed, but it is based on entirely different physical assumptions.

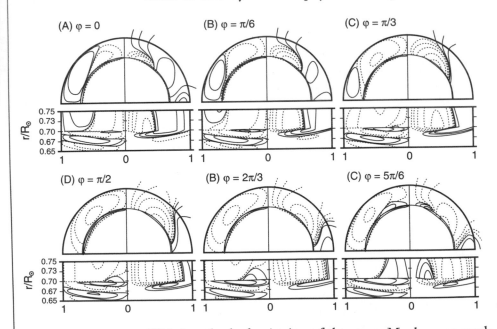

This is only the beginning of the story. Much more work needs to be done not only to clarify the behavior of the sun near the critical interface, but also to begin anew to build dynamic models. Time will tell whether we can ever build a realistic model that has predictive power.

Meanwhile astronomers are searching for other clues to the design of the solar cycle by studying the cycles (and lack of cycles) among other stars.

CYCLES BY THE DOZEN

Way back in the early 1960s, Olin Wilson, a staff scientist at the Mt. Wilson Observatory, began a long-term study of the chromospheres of cool stars. He decided to measure the brightness of two spectral lines of ionized calcium (at 393.3 and 396.8 nm) that are very deep and wide in the spectra of cool stars. From the work of solar observers, Wilson knew that the centers of these two lines are especially bright in active regions, so they were well suited to follow stellar activity.

Wilson was a patient man, content to make careful, systematic observations of a handful of stars, month after month, for several years. In those days, the 100-inch telescope at Mt. Wilson was underused so that Wilson had

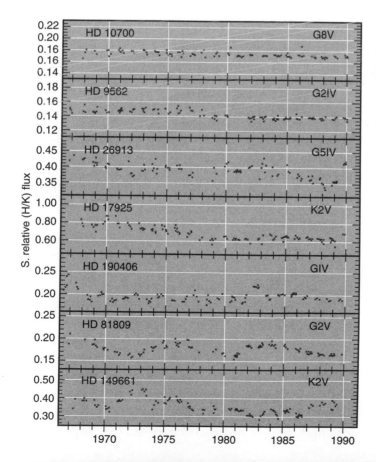

Fig. 13.7 A sample of some of Olin Wilson's stellar cycle observations. The graphs show how the brightness of the cores of two strong spectral lines of ionized calcium vary. (Courtesy of R. W. Noyes)

plenty of opportunity. He eventually expanded his sample to ninety-one stars.

After seven years of patient labor, Wilson began to see some interesting results. Some stars showed no activity whatsoever, some varied erratically by 10%, some showed a steady decline, others gave hints of cyclic behavior. Wilson persisted, and after twelve years he was able to exhibit the whole range of behavior (Figure 13.7).

The only stars that had nice clean cyclic variations were relatively old, like the sun. Arthur Vaughn, Wilson's colleague, determined that these stars, like the sun, have long rotation periods, exceeding twenty days. But not all slow rotators are cyclical; some with rotation periods as large as forty days show no variation whatever. In general the youngest stars rotate fastest (in a period as short as three days) and have the biggest swings of activity.

From these results it's clear that stellar rotation period and activity are connected somehow. The most likely reason of course is that Wilson's stars, like the sun, generate magnetic fields through dynamo action, in which rotation (the omega effect) is a prime factor. But why should a star rotate slower as it ages?

The accepted explanation is that the old stars were not only more active in their youth but had much stronger winds. A wind carries off mass and angular momentum (spin), so that over billions of years it gradually brakes the rotation of a star's convection zone. For example, at an age of 70 million years the sun (a mere babe) may have rotated in 3 days, and at an age of 600 million years, in 10 days. Now in its middle age of 4.6 billion years, its equator rotates in 27 days.

Dynamo Models of Stars

Wilson retired in 1974, and died in 1994. His pioneering observations of almost one hundred stars were continued until recently at Mt. Wilson Observatory by a group of dedicated astronomers. From their work, we now have a 25-year record of activity for these stars. About a third of them turn out to have stable cycles of activity, and like the sun, also show long-term variations in the peaks and periods of their cycles.

A number of theorists are beginning to build numerical dynamo models in order to understand the range of variation one sees among these stars, and therefore to place the sun's behavior in context. We'll examine the work of one of these mathematical physicists, Nigel Weiss, and his colleagues.

Weiss is a soft-spoken Englishman, a professor at Cambridge University. His specialty is the hydrodynamics of convection in magnetic fields. With his students and colleagues he has recently been applying concepts in nonlinear mechanics, better known as "chaos theory," to understand stellar cycles.

As we have seen, it is easy enough to construct a "kinematic" dynamo model for a magnetic star, in which the motions of rotation (the omega effect) and helical convection (the alpha effect) are simply assumed. When the parameters of the model are chosen suitably, it will churn out strictly periodic oscillations of the toroidal magnetic field, somewhat similar to the record of sunspot numbers.

Weiss and company want to advance beyond this stage to try to reproduce the more subtle features of the sunspot record such as the long-term variations in the peaks and periods of stable cycles. More important they want to understand why some of Wilson's sun-like stars are totally inactive, why some have stable activity cycles while others behave erratically, and finally why some stars' activity seems to be winding down steadily over decades.

Their strategy is to investigate *nonlinear* kinematic dynamos, in which the magnetic fields generated by rotation and convection are no longer passive. Instead, when they are strong enough, they can *change the very motions that generated them*. The changed motions then generate different magnetic fields, and so on.

Weiss and associates have demonstrated that a star can wobble between two distinct states, like a manic-depressive. In the first state, the motions of the star are stable but its magnetic field is unstable so that some kind of magnetic cycle is produced. In the second state the magnetic field is stable but the star's motions are unstable.

Once it starts up, a nonlinear star spontaneously drifts from one state to another over a long time. As you watch, you will see the first state, cyclic behavior with long-term variations in the peaks and lengths of the cycles. Figure 13.8 is one example, chosen to mimic the sunspot record. Notice

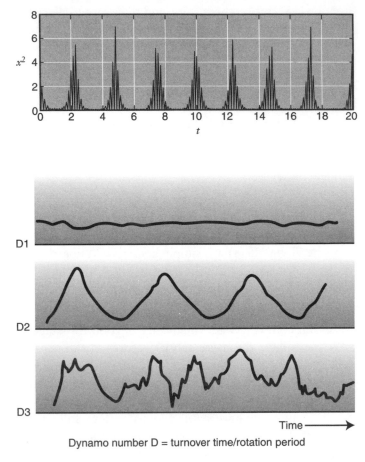

Fig. 13.8 A nonlinear model of the solar cycle produces a series of sunspot counts. (Courtesy of N. Weiss)

Fig.13.9 As the dynamo number is increased slowly, from D1 to D3, a star will proceed from no activity, to a regular cycle, to chaotic activity. (after N. Weiss et al., 1998)

Dynamo number D = turnover time/rotation period

how the peaks rise and fall over fourteen cycles. But in this toy model, the shapes of the cycles repeat rather monotonously. The real sun is even more chaotic than this model.

After many cycles, a star spontaneously shifts to the second state, turns off its magnetic activity, and begins slowly to change its internal motions. We'll come back to that state in the next chapter when we discuss the sun and the earth's climate.

But let's continue with Weiss's models of Wilson's stars. A star's magnetic behavior is governed by two critical periods: the time τ for a convective cell to turn over and the equatorial rotation period P. The ratio $D = \tau / P$ is the so-called dynamo number. Dramatic changes occur at critical values of D (endnote 13.5).

If you compare Figures 13.8 and 13.9, you see that these nonlinear dynamo models of Weiss and associates can reproduce many of the features of Wilson's stars merely by changing D. There is clearly much more to be done, since such models are merely kinematic (their motions were chosen, not predicted from basic physical principles), but they are clearly on the right track.

Weiss and his colleagues have been able to extend their models to interpret another feature of solar and stellar cycles, the occasional appearance of long intervals in which the star is completely inactive. These results have direct bearing on the earth's climate, to which we turn next.

Chapter 14

THE SUN AND CLIMATE

On Christmas Day, 1690, snow flurries were drifting over the Italian Riviera, and in normally sunny Rome, English woolens were in great demand. For the first time in a generation, children in London were ice-skating on the Thames. In Amsterdam the canals were frozen from top to bottom and the snow lay deep in Paris.

Europe had never seen a winter like this one. In fact, winters had been especially severe for as long as the graybeards could remember. The summers were also unusually cold. Looking back on this period, meteorologists named it the "Little Ice Age."

Two hundred years later, E. Walter Maunder, a British astronomer who was captivated by the idea of canals on Mars and who wrote a book titled "Astronomy of the Bible," was poring over old sunspot records. He stumbled on a curiosity.

For over seventy years, from about 1645 to 1715, hardly any sunspots were visible. In effect, six sunspot cycles were entirely missing! Gustav Spörer, a German astronomer, had noticed the same thing a few years earlier, but hadn't made much of it. Maunder, however, had the advantage of knowing about David Douglas's dating of tree rings.

By matching annual growth rings in trees of various ages, Douglas was able to assemble a long time-series. He discovered an eleven-year cycle in ring widths, identified it with the period of the sunspot cycle, and suggested that the sun influences rainfall, the dominant factor in tree growth in the American southwest. Maunder then speculated that the sun is a strong influence on other aspects of climate and that an absence of sunspots is the key marker. Maunder knew nothing about the severe winters of the late seventeenth century, but in 1976, Jack Eddy put the two events together.

Eddy is a genial fellow, whose laid-back manner masks a sharp intelligence. He spent the early years of his career in Boulder, Colorado, as a fairly productive solar physicist, with an unusual interest in the history of science. He did his research and published papers, but he was not noticeably one of the luminaries in the field, until he fell onto the Maunder Minimum.

At the time, he was working at the High Altitude Observatory, which was focused on the sun's possible influence on climate. Scientists there had analyzed the thickness of Douglas's tree rings and showed that certain parts of the American southwest experienced droughts in a 22-year cycle, at least for a few cycles.

Eddy's interest in history was the key factor in his discovery. He had read about both the Little Ice Age and the Maunder Minimum. He began to wonder whether the sun was responsible for Europe's terrible winters.

Eddy was cautious. His first task was to establish the reality of the Maunder Minimum beyond doubt. Wasn't it possible that astronomers at that time simply weren't interested in sunspots? To find out, Eddy dug out the original records at several major European observatories. He learned that several respected astronomers had persistently searched for sunspots and were puzzled by their absence. Other aspects of solar activity, like a bright corona, were also missing.

A variety of clues helped to convince Eddy that the spot cycle had essentially switched off for some seventy years. At the same time, incredibly cold weather had prevailed in Europe. There did seem to be a definite connection between the two events. To clinch the argument, Eddy pointed out that Europe's climate returned to normal once the sunspots reappeared, and has remained normal to this day. This single instance of a clear connection encouraged many scientists to search for further evidence and to establish the physical causes.

More recently, concern about greenhouse warming has motivated research into climate changes. The decade of the 1990s has been the warmest in this century and our consumption of fossil fuels seems to be the major cause. However, scientists have found remarkable changes in the past history of the earth's climate, long before the industrial age, and are still debating their causes. The sun's activity could be only a minor factor, but nobody is sure.

The experts do agree on the cause of the really long ice ages that lasted tens of thousands of years. According to current ideas, the amount of sunlight the earth receives varies because of very slight changes in three parameters of the earth's orbit, with periods of about 100,000, 41,000, and 23,000 years. But what causes the climate to change over 1,000, 100, or even 10 years?

To the average person it makes sense that the sun should influence our climate. However, many climatologists who study the evidence are not convinced that a physical connection exists. In fact it's difficult to explain just how the sun might affect the earth's atmosphere enough to make a difference. And yet new evidence for a solar influence continues to appear. In this chapter we'll follow the latest developments in this saga.

But first we'll finish Eddy's story.

LOOKING BACKWARD

Eddy was able to show that the Maunder Minimum was not unique. He was fortunate in obtaining a record of solar activity over the past nine thousand years that other scientists had reconstructed from measurements of an isotope of carbon ("carbon 14") in old tree rings (see Figure 14.1). He discovered that the Maunder Minimum appears in the correct place in this record and that several similar episodes of very low activity, each lasting for several decades, also appear in past centuries. Moreover, several peaks of very

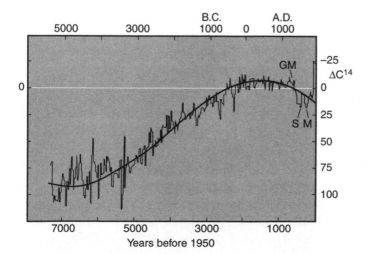

Fig. 14.1 The variation of the amount of carbon 14 in tree rings, dating back to 5000 B.P. The big sinusoidal swing is caused by a slow change in the earth's magnetic field. The wiggles are indications of solar activity. $\underline{M}$ is the Maunder Minimum; $\underline{S}$ is the Spörer Minimum; $\underline{G}$ is the Grand Maximum in the twelfth century. (P. E. Damon's data, courtesy of J. A. Eddy)

high activity corresponded with unusual warming trends. This additional evidence strengthened the case for a sun-climate connection and encouraged other scientists to look further.

Because the carbon 14 record is a key player in this subject, it is worth taking a moment to understand how it connects the sun and the earth (see endnote 14.1).

At the same time Willard Libby, a chemist at the University of Chicago, was refining his ^{14}C dating scheme, C. W. Furguson at the University of Arizona was matching annual growth rings in samples of the bristlecone pine tree. These small, hardy trees are found in the arid southwest and can live for as long as five thousand years. Furguson was able to construct a continuous series of rings reaching back nine thousand years. By counting the annual rings he could establish the exact year in which any ring grew.

Several scientists stepped forward at this time to compare the ^{14}C and tree ring dating schemes. When they dated the tree ring series with Libby's radiocarbon method, they found some discrepancies. Some samples contained more or less ^{14}C than the tree ring dates would predict. These "wiggles" in the ^{14}C record are shown in Figure 14.1. (The curve has been inverted to correspond to sunspot numbers.)

Notice the distinct dips in the record during the Maunder Minimum and the so-called Spörer Minimum in the fifteenth century. A peak, corresponding to enhanced solar activity, appears in the twelfth century. How can one explain this obvious correlation between amounts of carbon 14 and the number of sunspots? The answer lies in the way this isotope is formed (see endnote 14.2).

For the single instance of the Maunder Minimum we know that a peak in the ^{14}C record corresponds to a Little Ice Age, at least in Europe. Does this mean that each peak in the ^{14}C record also corresponds to a mini–ice age somewhere? That is an attractive working hypothesis, but how can one test it?

THE MINUET OF THE GLACIERS

Vast sheets of ice covered most of North America and Europe during the latest major Ice Age, which lasted a hundred thousand years. During the past ten thousand years the earth has warmed and the sheets have shrunken, but ancient glaciers have survived at high latitudes. Survived perhaps, but not without change. Most glaciers have retreated during the past century, some as much as a kilometer, but in the more distant past they advanced rapidly. Geologists have charted this minuet of the glaciers, in both hemispheres, over the past ten centuries. Figure 14.2 shows the record for each hemisphere separately and for the earth as a whole. The two hemispheres were obviously well synchronized, so that whatever caused the warming and cooling affected the entire globe uniformly. Could the sun be responsible?

In order to test this idea, two scientists at the University of East Anglia compared the glacial record with the wiggles in carbon 14. In Figure 14.3, the shaded bands in the lower curve indicate the peaks in carbon 14 which correspond to low sunspot activity and possibly also to cold periods on earth.

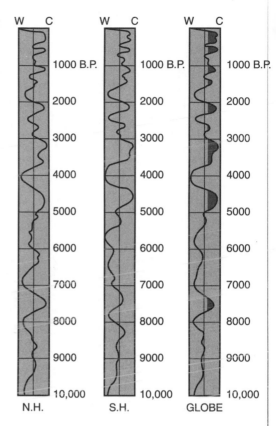

Fig. 14.2 Alpine glacier advances and retreats in the Northern Hemisphere (left column), Southern Hemisphere (middle), and over the globe (right). (Courtesy of T.M.L. Wigley and P. M. Kelly, Phil. Trans. R. Soc. A330, 547, 1990)

Some bands coincide nicely with glacial advances, but others don't. The probability that the coincidences occur by chance is only 3%. That is good but not outstanding. "Overall," the scientists wrote, "we consider these results to be highly suggestive of a ^{14}C-climate link but not ultimately convincing."

There's more than one way to skin a cat, however. In hot pursuit of climate changes, glaciologists have turned to the huge ice sheets left over from the last major Ice Age.

HISTORY WRITTEN IN THE ICE

Greenland and Antarctica are each covered with ice sheets several kilometers thick. Every year new snow falls, the previous year's accumulation is compressed, and a new layer of ice is added to the stack. In cold years the layers are thicker, in warmer years, thinner. So, like tree rings, the layers in ice sheets contain a long history of climate changes in the

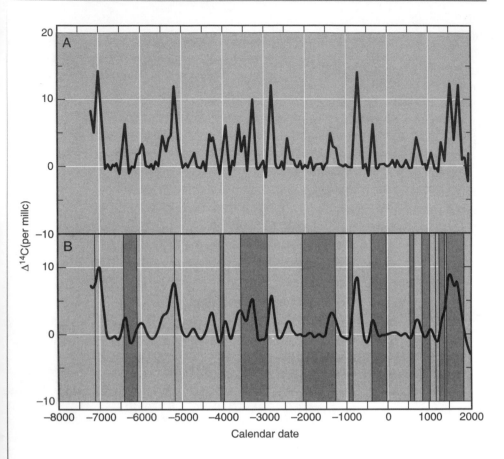

Fig.14.3 A comparison
of carbon 14 wiggles
and periods of cold
climate (shaded bands).
(Wigley and Kelly,
1990)

polar atmosphere.

During the past decade scientists have drilled down into these polar ice sheets to extract cores that reveal the annual layers. The thickness of the layers is only one clue to the climate, however. As we shall see, some very sensitive techniques have been developed that use radioactive isotopes in the ice to recover the air temperature at the time the ice formed.

In 1990, G. M. Raisbeck and his colleagues in France analyzed two ice cores obtained from Antarctica. One, covering the past one thousand years, was taken from the South Pole and the other, covering the past three thousand years, was extracted from a neighboring ice dome.

Instead of a ^{14}C record from tree rings, they used the amount of beryllium 10 (^{10}Be) contained in the ice itself to reconstruct past solar activity. This radioactive isotope is formed by cosmic ray impacts in the same way as ^{14}C. Then, as an indicator of polar air temperature, they measured the amount of deuterium ("heavy hydrogen," ^{2}H) in the *same* samples of ice. That procedure guarantees that they are able to compare solar activity and polar air temperature on the *same* date.

Deuterium is a good thermometer because water molecules containing deuterium evaporate more slowly than normal water molecules and the lower the air temperature, the bigger is the difference. The same is true for the heavy isotope of oxygen, ^{18}O. Therefore when snow falls at lower temperatures, it contains less of these heavy isotopes.

Figure 14.4 compares the wiggles of ^{10}Be and ^{2}H along the 3,000-year-long ice core. As you can see, the match between solar activity and polar air temperature is not especially close. In fact, a statistical test shows that there is at least a 10% chance the match could have occurred by chance

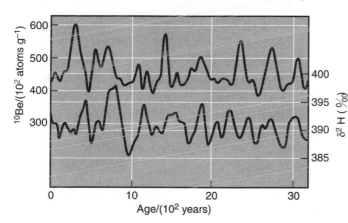

Fig.14.4 A comparison in Antarctic ice of beryllium 10 (indicating solar activity) and deuterium (indicating air temperature). (G. M. Raisbeck et al., Phil. Trans. R. Soc. A330, 463, 1990)

between completely unrelated curves. That isn't nearly good enough to satisfy the critics, so once again, these data don't say anything conclusive about the sun's influence on polar climate.

Perhaps we should relax our criteria of success. We have been assuming, implicitly, that the sun influences the climate equally everywhere over the earth. But after all, each continent and each ocean may create its own micro-climate. So perhaps the Antarctic is not as sensitive to solar forcing of climate as some other regions. It's worth looking elsewhere.

THE GREENLAND ICE CAP

For the past decade, two independent groups of scientists have been drilling deep into the ice over Greenland. In 1992, the European Greenland Ice Project (GRIP) extracted a continuous core 3,029 meters long, containing a record over 100,000 years in length. A year later, the Greenland Ice Sheet Project Two (GISP2) removed a core 3,053 meters long from a site some 30 kilometers from the GRIP site. Each layer in the cores contains gases and aerosols, trapped at the time of formation, that contain valuable clues to the history of Greenland's climate.

Perhaps the first thing one wants to learn is how the temperature varied over the past hundred centuries, since the end of the last major ice age. For this purpose, scientists have measured the small amounts of radioactive oxygen (^{18}O) that are trapped in the ice. Like deuterium (^2H), this isotope serves as a thermometer. The ratio of the concentrations of oxygen 18 to the stable isotope, oxygen 16, is a measure of the air temperature at the time the snow fell. An extremely sensitive technique (mass spectrometry) allows scientists to isolate a few atoms per gram of ice and from these to determine air temperature.

In 1995, scientists from the University of Washington Fourier-analyzed 11,500 years' worth of the GISP2 ice core, to see how the air temperature varied. Figure 14.5 shows the periods that show up in two pieces of the record. The top

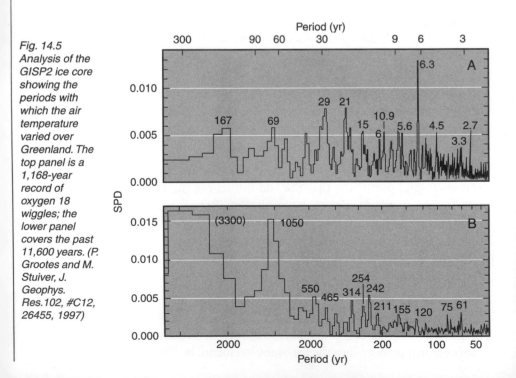

Fig. 14.5 Analysis of the GISP2 ice core showing the periods with which the air temperature varied over Greenland. The top panel is a 1,168-year record of oxygen 18 wiggles; the lower panel covers the past 11,600 years. (P. Grootes and M. Stuiver, J. Geophys. Res.102, #C12, 26455, 1997)

panel shows the periods active in the years between A.D. 818 and 1985, while the bottom panel covers the whole record back to 11,500 B.P. This longer record was first smoothed to exclude periods shorter than 20 years.

As you can see, the air temperature varied with a great number of discrete periods. Of these, the 10.9- and 21-year peaks (and possibly the one at 211 years) suggest standard solar cycles but obviously there is much more going on. Also, one might expect the well-known 88-year period of sunspot numbers (the "Gleissberg" period) to show up in ^{18}O, but it doesn't.

The team pressed harder on the data to search for a possible eleven-year solar cycle influence on Greenland's climate. As you can see in endnote 14.3, they thought they found one.

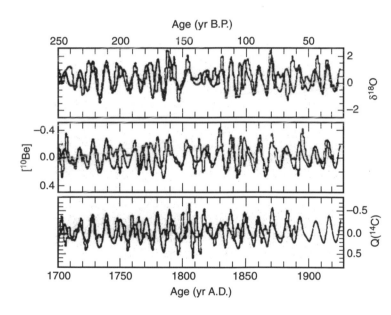

Fig.14.6 GISP2 ice core results. A comparison of air temperature (oxygen 18) and solar activity (beryllium 10 and carbon 14) with sunspot numbers (heavy line). The curves match up nicely, but only after some arbitrary shifts in time. (M. Stuiver et al., Quaternary Research 44, 341, 1995)

Let us accept their conclusions for the moment. What about the other long ice core, obtained by the GRIP project? Does it confirm an eleven-year cycle of climate in Greenland? Well, yes and no.

A team of Danish scientists looked for periodicities in their ^{18}O record, covering the past 917 years. They found three distinct periods of 61.6, 19.4, and 11.6 years. Now 19.4 is not 22, nor is 11.6 exactly 11, so how much latitude can one allow before one can claim a solar connection? Not much, the Danes would say. They studied how the periods change as more or less data is included in the analysis and concluded that the discrepancies are too large. One might quibble with them since, as we learned earlier, the so-called "11-year" sunspot cycle really varies from 8 to 14 years, but this group states unequivocally that a solar connection is not proven.

So we have a controversy, with conflicting results. Bear in mind that the two ice cores were taken from sites only 30 km apart in Greenland. For depths corresponding in age greater than 10,000 years the ^{18}O wiggles in the two cores correlate beautifully, but in the very interval (the past 9,000 years) where ^{14}C could test a solar influence on climate, the two cores differ.

The main point that comes out of these studies is that, in the past, the sun was at best only one factor, perhaps a very minor one, in the complicated variations of climate of Greenland, and presumably of every other place as well.

And yet the quest for a solar connection continues. Every now and then a new study reveals what appears to be a firm association between the sunspot cycle and the climate. We've mentioned the 22-year cycle of droughts in the southwest United States. Here are a few more recent examples. They are difficult to ignore, but even more difficult to explain.

WINDS AND WARMING

In the stratosphere above the North Pole the wind normally blows from west to east, circling the Pole as a cold whirlpool. In certain winters, however, the winds reverse direction and the Arctic becomes warmer than, say, Maine. The warming trend lasts about a month and spreads horizontally to the tropics. Karin Labitzke, a scientist at the Free University of Berlin, and her coworker Henry van Loon, discovered a connection between these warming events, the eleven-year sunspot cycle, and a curious oscillation of the winds in the tropical stratosphere.

Figure 14.7 shows their original results. In the top panel the polar air temperature at an altitude of 22 km is compared to the sun's radio emission at 10.7 cm. (This emission is a good proxy for sunspot numbers and is a convenient measure of solar activity.) The two curves overlap in a jumble, without any obvious correlation.

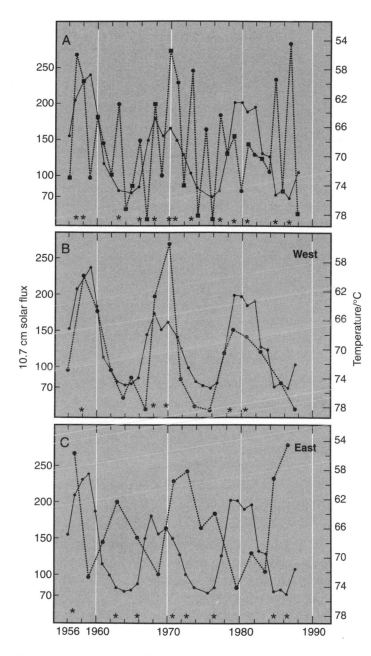

Fig.14.7 North pole temperatures compared with a proxy for solar activity (10.7 cm radio radiation). In the top panel the two variables are compared directly; asterisks show major mid-winter warmings. In the middle and bottom panels the data are sorted according to the direction of the wind in the tropical stratosphere (west in the middle, east in the bottom panel). Now a nice correlation appears. (K. Labitzke and H. van Loon, Phil. Trans. R. Soc. A330, 577, 1990)

However, in the middle and bottom panels the temperature data have been sorted into two bins, according to whether the stratospheric winds in the *tropics* are blowing from the west (the middle panel) or from the east (the bottom panel). The tropical stratospheric winds reverse direction about every two years in a so-called "quasi-biennial oscillation" or QBO.

The point of the figure is that the polar midwinter warming events occur at sunspot *maximums* when the tropical wind is westerly and occur at sunspot *minimums* when the tropical wind is easterly! The correlation is now obvious to the eye and statistical tests confirm that it is highly significant.

Labitzke and van Loon demonstrated a similar effect at the South Pole and more important, that the warming extends right down to the earth's surface. In the west phase of the QBO, North America and the Atlantic warm up, while in the east phase it is the northern Pacific. In later work, the two scientists showed that the temperatures from the surface to the stratosphere, at several places around the globe, vary in step with the eleven-year cycle, *independently of the phase of the QBO.*

Although these results greatly strengthened their argument, they were criticized on at least two counts. First, the effect was demonstrated only for three eleven-year cycles and many other promising associations were known to break down after a few cycles. Secondly, no physical mechanism linking the sun and the atmosphere was proposed to explain the effect. Quite possibly, the critics said, the atmosphere could produce these effects without any solar forcing. The critics would await more convincing evidence.

It was not long in coming.

THE LONGER CYCLES

In 1987, George Reid, a climatologist at the National Center for Atmospheric Research in Colorado, made an inspired guess. He realized that the oceans can sop up a lot of heat before their temperatures change by any significant amount. Therefore, he thought, if the sun does somehow affect the sea's temperature, its influence must persist for longer than the classical 11-year cycle. He decided to compare changes in the temperature of the sea surface, averaged over the whole globe, to the longer-term variations of sunspot numbers. As we learned earlier, the 11-year cycle is mixed with longer periods, such as the 88-year "Gleissberg" cycle. To his delight, Reid found a striking similarity between the 88-year cycle and the sea temperature.

Two Danish scientists picked up this clue in 1991. They knew that the so-called 11-year sunspot cycle actually varies in length from about 8 to 14 years, with a Gleissberg cycle of 80 to 90 years. Short cycles correspond to higher peaks of activity and vice versa. So they compared the *length* of the solar cycle with changes in the surface land temperature over the Northern Hemisphere. Figure 14.8 shows the nice agreement between these two variables, suggesting a possible causal connection (see more in endnote 14.4).

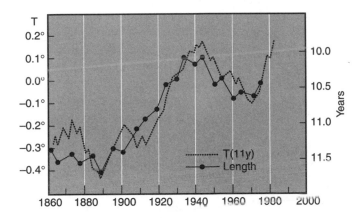

Fig.14.8 Land temperatures in the Northern Hemisphere compared to the actual length of the 11-year sunspot cycle. (K. Lassen and E. Friis-Christensen, J. Atmos. Terr. Phys. 57, #8, 815, 1997)

All well and good, but a correlation doesn't prove a causal connection. "Give us an explanation in terms of *physics*," the skeptics said.

THE SUN'S BRIGHTNESS?

The most obvious direct solar influence on earth is, of course, the total amount of sunlight the earth receives. As we learned earlier, ACRIM measurements of sunlight from a series of spacecraft, beginning in 1980, revealed a 0.1% variation of the sun's luminosity, in step with the 11-year solar cycle. While dark sunspots dim the sun slightly, the chromosphere above magnetic regions brightens it slightly. The two effects nearly balance, but a small difference of less than 0.1% remains. Is such a small change, over such a short interval, sufficient to explain the correlations in climate that have been observed?

One way to answer this question is to exercise the huge computer models of the earth's atmosphere that climatologists have developed. The best models are three-dimensional and time-dependent. They take the sun's known spectrum as input, and treat the atmosphere as a heat engine, taking into account the complicated photo-chemistry of the upper atmosphere. The latest models are helping to explain the patterns of winds, ocean currents, and precipitation that are observed over long time spans. But even the best models are far from perfect. They all have weak spots, especially the messy physics at the sea-air surface, and in the formation of clouds.

Nevertheless they are useful in estimating the effect on the climate of a varying "solar constant" of sunlight. Although the different models yield slightly different results, they are unanimous in claiming that an 11-year, 0.1% variation is too small to explain the original results of Labitzke and van Loon, if for no other reason than that the Arctic Ocean would damp out the variation. Either some other link to the sun must be found, or this effect may have nothing to do with the sun.

On the other hand, the sun's tiny changes in brightness might explain the Labitzke–van Loon results over the *continents*, because the land heats up and cools down more rapidly than the oceans.

Over time spans longer than eleven years, the picture is a little more encouraging. George Reid estimated the change in the sun's brightness that would be needed to explain his results on global sea temperatures. Several tenths of a percent, over a period of decades to centuries, would be sufficient.

There is no *direct* evidence that the sun's brightness changes this much over long intervals, simply because ACRIM observations only began in 1980. But two lines of research make such long-term changes believable.

The first is the observation of long intervals of reduced brightness in some solar-type stars. As we saw, the Mt. Wilson observers have good examples of this type of behavior (see Figure 13.7).

The second clue comes from the computer models of Nigel Weiss and colleagues. They have shown that nonlinear stellar dynamos can spontaneously switch into states of low or minimum activity that last for decades—a Maunder Minimum in effect. We still don't know precisely what caused the Little Ice Age in Europe, but it might have been a dimming of the sun that was associated with low activity, and lasted seventy years instead of the usual eleven.

YET ANOTHER IDEA

In the search for a way in which the sun can modulate the climate, researchers have explored the idea of *amplification*.

Suppose a relatively small change at the sun somehow stimulates a much larger response in the earth's atmosphere. (The last straw on a camel's back, so to speak.) Several schemes along these lines have been proposed recently.

Another pair of clever Danes uncovered a surprising relationship between global cloud cover, as observed by satellites, and the cosmic ray flux entering the atmosphere. As we learned earlier, the magnetic field of the solar wind shields the earth from cosmic rays. As the field strength varies during the eleven year sunspot cycle, the cosmic ray flux reaching ground level varies in step, changing by a factor of a hundred.

The Danes found that the amount of cloud cover over the whole earth rose by 2% during the last solar minimum (1986) and fell by about 1% during solar maximum. These are relatively small changes, but they could have had important consequences for the climate.

What caused these changes is still uncertain. The Danes suggested that free charges or ions, produced by the cosmic rays in the stratosphere and lower atmosphere, collect on aerosol particles. Charged particles are known to be more efficient in condensing water droplets than uncharged particles. So the formation of clouds may depend in part on free ions, whose numbers vary with the solar cycle.

Another example of amplification involves the ozone in the stratosphere (see endnote 14.5). This study, among others, reveals the importance of a dynamic coupling between the stratosphere and troposphere, and the sensitivity of the whole atmosphere to tiny changes in the sun's radiation. Perhaps, at last, we are beginning to understand the physics that links the sun and the earth. The problem is immensely complicated by the inherent nonlinearity of the earth's climate system, in which the different layers of the atmosphere interact with each other and with the oceans. The possibility remains that the majority of eleven-year oscillations have little to do with the sun, and are inherent in the complicated gyrations of the atmosphere alone.

Chapter 15

A FEW LAST WORDS: WHERE DO WE GO FROM HERE?

 he missing neutrino problem seems much closer to solution now that physicists recognize that neutrinos may change in flight. But three really big questions in solar astronomy that have been unresolved for decades remain to be answered. How is the corona heated? How is flare energy stored and released? How is the fast solar wind accelerated? In addition, such questions as the origins of coronal mass ejections and the solar cycle have moved to center stage.

The prospects for definitive answers to some of these questions are good. We are already getting clues from the instruments aboard SOHO and TRACE on coronal heating and the origin of fast wind. As for flares, YOHKOH has helped considerably, but important issues concerning the most energetic radiation remain to be settled.

In 2001, NASA plans to launch the High Energy Solar Spectroscopic Imager (HESSI), a satellite observatory instrumented to study charged particle acceleration and energy release in flares. For the first time it will be possible to obtain images in hard X-rays and gamma rays, with a resolution of about 700 km on the sun. A spectrometer will cover the entire photon energy range, up to 20 million electron volts. This powerful array of instruments should go a long way toward revealing how electrons and ions are accelerated to relativistic energy in flares.

Also in 2001, a Japanese-American consortium expects to launch SOLAR B, another solar space observatory. It is aimed at a detailed study of solar magnetic fields and their role in heating the corona. This satellite will carry a vector magnetograph and several ultraviolet telescopes. A strong research group has been gathered to analyze the data.

Not all the action will occur in space. The GONG network is being upgraded with larger detectors in order to study intimate details of the sun's interior with sound waves. We can soon expect progress in understanding how sunspots look under the surface and possibly how active regions rise to the surface. More exciting perhaps are the prospects of pinning down just where and how the solar dynamo generates magnetic fields and why the solar cycle varies in length.

For some time solar astronomers have needed larger ground-based telescopes to make more precise observations, especially of very small, very dynamic magnetic loops. The 90 cm French telescope THEMIS will soon be commissioned in the

Canary Islands for just that purpose. Very soon, the Swedish Vacuum Solar Telescope in the Canary Islands will also have a new optical system that will fully exploit the excellent observing conditions there.

In the United States, solar astronomers are hoping to build an Advanced Technology Solar Telescope, which would have a mirror 4 meters in diameter and would incorporate the latest technology of on-line image restoration. It should be able to resolve the tiniest details of magnetic reconnection in flares, and the dynamic behavior of emerging magnetic fields on the quiet sun.

Not everyone is interested in tiny features on the sun, however. As we have seen, the largest magnetic structures are organized in remarkable patterns that change systematically during the cycle. To follow this slow evolution, the United States and its foreign partners are building a network of identical stations (Synoptic Optical Long-term Investigations of the Sun, SOLIS). Each station will be equipped with imaging magnetometers and other special telescopes. The data will be archived for the use of all interested researchers.

If the past is any guide to the future, astronomers will find some pleasant surprises waiting for them when they turn on these marvelous instruments. Research does not advance only by pre-programmed observations. Every new tool invariably opens new opportunities for discovery. Nobody can guess what they may be just now.

Nor can anyone guess what some clever young theorist will propose next. In solar astronomy, observations have generally guided theory, but with the art of computer simulation advancing so rapidly, theory may take the lead.

All in all, the next decade in solar astrophysics promises to be an exciting one!

NOTES

Chapter 2

THE SECRET HEART OF THE SUN

2.1 The Proton-Proton Chain

The first step, which we saw earlier, is the collision of two protons, with the formation of a deuteron, which consists of a proton and a neutron. Neutrons and protons have nearly the same masses and one can change to the other by gaining or losing a unit positive electric charge. In this reaction the positive charge of one proton is carried off by a new particle, a positron, which is a positively charged electron. In addition, a neutrino, a particle nearly without mass or charge, is created. We can write the reaction as

$$^1H + {}^1H \rightarrow {}^2D + e^+ + \nu_e.$$

The symbols correspond to ^1H for a proton (the nucleus of the hydrogen atom), ^2D for the deuteron (with a mass of two protons), e^+ for the positron, and ν_e for the neutrino. The three final particles carry away the released energy in the form of kinetic energy, which amounts to 1.4 million electron-volts or Mev. The neutrinos have a maximum energy of 420 kilovolts.

Once a deuteron is formed it may combine with another proton to form a light-weight version ("isotope") of helium, ^3He, with mass of three protons, as follows:

$$^2D + {}^1H \rightarrow {}^3He + \gamma.$$

The gamma γ in the above equation indicates that a gamma ray (a hard X-ray) is released, with an energy of 5.5 Mev. From here on there are two paths to complete of the chain. In the first, two ^3He combine to form the standard form of helium, ^4He, with a mass of four protons.

$$^3He + {}^3He \rightarrow {}^4He + {}^1H + {}^1H.$$

Two protons are ejected and the three final particles carry away kinetic energy of 12.9 Mev.

In the second path, a beryllium nucleus ^7Be is formed which picks up an electron and converts to a lithium nucleus ^7Li and then to ^8Be, as follows:

$$^3He + {}^4He \rightarrow {}^7Be + \gamma$$

$$^7Be + e^- \rightarrow {}^7Li + \nu_e$$

$$^7Li + {}^1H \rightarrow {}^8Be + \gamma + {}^4He + {}^4He.$$

Alternatively, the chain may end with the formation of a boron nucleus, ^8Be:

$$^7Be + {}^1H \rightarrow {}^8Be + \gamma$$

$$^8Be \rightarrow e^+ + \nu_e + {}^4He + {}^4He.$$

The end result of all these reactions is to fuse four protons into a helium nucleus, ^4He, with the release of two positrons, $e+$, two neutrinos, νe, and up to 26.7

megavolts of energy in the form of gamma rays and the kinetic energy of created particles. The latter is redistributed by collisions very quickly among all neighboring particles and serves to maintain the local temperature.

The Carbon Cycle

In 1939 Hans Bethe devised a second possible chain reaction to convert hydrogen to helium, the carbon-nitrogen-oxygen chain or CNO. It uses carbon nuclei as catalysts to produce the same amount of energy per helium nucleus, 26.7 Mev, as in the proton-proton chain. Although the CNO chain produces only about 1.5% of the solar luminosity, it does contribute substantially to the flux of neutrinos with energies below about 2 Mev.

The principal reactions are as follows:

$$^{12}C + {}^1H \rightarrow {}^{13}N + \gamma$$

$$^{13}N \rightarrow {}^{13}C + e^+ + \nu_e$$

$$^{13}C + {}^1H \rightarrow {}^{14}N + \gamma$$

$$^{14}N + {}^1H \rightarrow {}^{15}O + \gamma$$

$$^{15}O \text{ Æ } {}^{15}N + e^+ + \nu_e$$

$$^{15}N + {}^1H \rightarrow {}^{12}C + {}^4He.$$

Notice that the initial ^{12}C nucleus goes through a series of transmutations, increasing in mass with the addition of protons, but ultimately returns to a ^{12}C.

To create nuclei heavier than helium requires quite different chain reactions, and these proceed efficiently only in stellar cores much hotter than the sun's. Thus the fact that the sun contains such heavy metals as iron, means that our sun is a second- or third-generation star. The cloud of hydrogen in which it formed must have been previously enriched by the contents of the core of a very hot star, which exploded as a supernova. This is the majestic life cycle of stars—to process elements in their cores and to pass on their legacy to a new generation.

2.2 The Homestake Neutrino Experiment

Every few days, a *single* nucleus of chlorine is converted to a radioactive nucleus of argon, ^{37}A, by the capture of a solar neutrino. About once a month, Davis flushes his tank with helium gas, which entrains the radioactive argon gas. Then he passes the helium and argon mixture through a cold charcoal trap, which freezes out the argon gas. The ^{37}A atoms eventually capture a free electron from their surroundings to reform ^{37}Cl atoms in an excited energy state. Those atoms then decay to normal unexcited ^{37}Cl atoms by emitting a fast electron. By counting the electrons with a Geiger counter, Davis can determine the number of original ^{37}A atoms and thus the number of solar neutrinos he has captured.

To check on the efficiency of his collection method, Davis introduces a little ordinary argon, ^{38}A, with the helium and measures the amount he recovers. In this way he can show he has recovered at least 90% of the radioactive ^{37}A.

If the sun were the only source of neutrinos, Davis's work would be a lot simpler. Unfortunately, energetic cosmic rays, colliding with the upper atmosphere, produce showers of secondary particles that can decay and produce neutrinos in the same energy range (above 800 kilovolts) Davis tries to measure. That is the reason he buried his tank in a mine—to shield it from cosmic ray showers. But the surrounding rocks in the mine also emit neutrinos as their uranium and thorium constituents decay radioactively. Their effect must be carefully calibrated and removed from his data.

2.3 Two Second-Generation Neutrino Experiments

Both GALLEX and SAGE have a threshold energy for solar neutrinos of only 0.223 Mev. The GALLEX experiment is a European effort, with American and Israeli collaboration. Thirty tons of gallium is contained in the form of a liquid, gallium chloride. The second experiment, SAGE, is a Soviet-American collaboration. Sixty tons of metallic gallium serve as the target in this experiment.

In both experiments an electron neutrino ν_e and a gallium 71 nucleus (^{71}Ga) combine to form a radioactive germanium 71 nucleus (^{71}Ge). The germanium nuclei can be collected by radiochemical techniques similar to Davis's, and yield a direct measure of the low-energy solar neutrinos.

2.4 Kamiokande II

The Kamiokande II experiment is also situated deep in a mine. It uses 680 tons of water as the primary target for solar neutrinos. An incident solar neutrino may, rarely, collide with an electron in the water molecules. If the neutrino energy exceeds 9 Mev the electron will have sufficient energy to generate Cerenkov radiation in the water. This deep blue light is emitted by an electron that travels faster than the speed of light in water, which is about 220,000 km/s. Cerenkov radiation is an optical analogy to the shock wave generated in air by a supersonic plane.

Cerenkov radiation is detected in the Kamiokande experiment by a layer of huge sensitive photomultiplier tubes that surround the water tank. Fast electronics enable the experimenters to determine the direction and energy of the colliding neutrino from the pattern of emitted light. As in all neutrino experiments, great care is necessary to evaluate the background radiation due to radioactive elements in the surrounding rocks and in the experiment's water.

Solar neutrinos with energy exceeding 9 Mev are produced primarily by the decay of ^{8}B nuclei in the proton-proton chain (see Figure 2.2). *Although the Kamiokande experiment detects only the most energetic solar neutrinos, its ability to detect all flavors gives it a unique advantage.*

2.5 Third-Generation Neutrino Detectors

The Sudbury Neutrino Observatory (SNO) was built by a UK-Canadian team in a deep nickel mine in Ontario, Canada. It uses 1,000 tons of heavy water (in which the protons are replaced by deuterons) as a target. Unlike in the Kamiokande experiment, incoming neutrinos scatter off the deuterons and produce electrons as a detectable by-product. The SNO has the crucial ability to distinguish electron neutrinos from other types in the decay of ^8B.

Borexino is a second detector, the product of a collaboration among Germans, Italians, Russians, and Americans. It is located in a deep laboratory about 100 km from Rome. It was designed specifically to measure the actual flux of neutrinos released in the formation of beryllium 7 (^7Be) and so check up on the puzzling results from GALLEX and SAGE. Instead of water, this machine uses a liquid that scintillates when a fast charged particle passes through it, including muon or tau neutrinos. So even if the electron neutrinos released in the formation of ^7Be change flavors in transit to the earth, they will show up as flashes of light in the Borexino machine.

The third detector is the Super-Kamiokande, located 200 km west of Tokyo. It is another water-Cerenkov machine but a huge one. With a target of 50,000 tons of water, it captures neutrinos with energies above 5 Mev at a hundred times the rate of its predecessor, Kamiokande. Its greatest virtue, however, is its ability to distinguish among all three flavors of neutrino.

Chapter 3

THE DEEP INTERIOR

3.1 The Perfect Gas Law

Strictly speaking, this law applies to very dilute gases and it's surprising that it also applies in the extreme conditions in the sun. The reason is that even at the high plasma density, the average distance between electrons is much larger than their diameters, so the plasma qualifies as a "dilute gas." Of course, the particles in the plasma carry electrical charges and this affects its compressibility, but these are relatively small effects. The perfect gas law works quite well as the "equation of state" of the plasma.

However, a small correction is necessary. The *total* pressure in the sun's radiative zone is composed of two parts: the pressure of the plasma and *radiation* pressure. After all, the intense blizzard of light that bombards us also exerts a pressure: on us and on the plasma all around us. In the sun this radiation pressure is never more than about 5% of the total but in stars much more massive and luminous than the sun this radiation pressure turns out to be comparable to the plasma pressure.

3.2 The Sun's Cooling Time

You can get a quick idea of how much heat is stored in the sun by asking how long the sun could continue to lose energy at its present rate if its energy generation in the core were suddenly turned off. The sun has a certain "luminosity," which measures the rate at which it loses heat. The time for it to cool must equal the amount of stored heat divided by the rate of heat loss. Now, the amount of stored heat depends on three things: on the total mass of the sun, on its mean temperature, and on the amount of heat it takes to raise the temperature of one gram by one kelvin (the "specific heat" of the plasma).

So the cooling time of the sun would equal this amount of stored energy divided by the luminosity. Using estimates of all the quantities (see the table in chapter 1), you would find a cooling time of several million years. Therefore the net amount of energy flowing through any blob in a year is roughly a millionth of the amount already present in the volume. Consequently, the temperature gradient across each blob is very small.

3.3 Radiative and Convective Processes

How can we understand Schwarzschild's result? Why should radiative energy tranport be more efficient than convection?

The reason has to do with the distances that photons and electrons can travel in the dense solar interior. A photon travels freely until it collides with an electron or a heavy ion. In traveling this "mean free path" the photon transports a small bit of energy. This is the essence of radiative transport.

In convective transport, heat must be exchanged from a hot to a cooler blob of plasma and it is the mobile electrons that carry this heat, in the form of their kinetic energy. But the electrons are electrically charged and their mutual repulsion impedes their motion. As a result, the average distance an electron can travel before it suffers a collision (its "mean free path") is shorter than that of a photon. The net result is that photons are more efficient carriers of energy.

3.4 The Probability Approach to Particle Velocities

Maxwell and Boltzmann derived their velocity distribution in a very curious fashion, without regard to the collisions we just spoke of. Instead, they asked how many ways can you divide a fixed quantity of energy E among N distinguishable particles.

They imagined distributing the particles among a set of bins (see Figure 3.6, top) that lie in a row. Each bin corresponds to a small range of kinetic energy. If a particle is placed in a bin, it acquires the energy appropriate to that bin. The one on the far left corresponds, let us say, to 0.0–0.1 kilovolts, the one to its right 0.1–0.2 kilovolts, etc. Then they assigned each particle to an energy bin, being careful not to exceed the total amount of available energy E. Their first attempt might look like Figure 3.6, top. But this particular division of the total energy looks the same if any two particles are interchanged. Therefore this arrangement can be formed in a large number of ways, say, A_1.

Then they tried a different arrangement (Figure 3.6, bottom) and counted the number of ways (A_2) *it* can appear, and so on. When they then asked (mathematically) which arrangement can appear the largest number of ways, after fixing both the total amount of available kinetic energy and the total number of particles, they found the distribution shown in Figure 3.5. It is the *most probable energy distribution* such a collection of particles can adopt.

This implies that despite momentary fluctuations in the distribution of particle energies (too many fast or slow particles) the distribution will return spontaneously to a Maxwell-Boltzmann in a short time. Collisions among the particles guarantee that happening.

3.5 How Particles and Radiation Interact

This is precisely the problem that Max Planck solved theoretically in 1896. He examined the behavior of a collection of oscillating electrons inside an isothermal enclosure. The electrons emit and absorb electromagnetic waves ("light waves") as they accelerate in their sinusoidal oscillations.

He chose to consider these oscillating electrons instead of real atoms because he knew that in thermodynamic equilibrium *all* systems capable of absorbing and emitting radiation must produce the same spectrum of radiation. So he chose the simple electron oscillator for his model of "matter," because its emission and absorption could be calculated precisely. (And incidentally, more easily than that of a Maxwellian electron gas.)

Now, an oscillating electron loses kinetic energy as it emits light. In an enclosure with perfectly reflecting walls, however, the electron will soon *absorb* exactly enough energy from the surrounding radiation to make up its losses. Similarly the radiation field loses energy when an oscillator absorbs some, but this loss is exactly balanced by the oscillator's emission. In this way the kinetic energy of the oscillator and the energy density of radiation at the oscillator's frequency are both maintained indefinitely. The oscillators come into thermodynamic equilibrium with the field of radiation. Planck's problem was to find the spectrum of that radiation.

He began by calculating the time-averaged spectrum of a single oscillator as it both emits and absorbs light. Figure 3.7 shows the result. The oscillator no longer radiates at its single "natural" frequency, v_0, but over a range of frequencies. Planck found the density of radiative energy (say, the number of calories per cubic centimeter) produced at all frequencies by such a system of oscillators is

$$U = 8\pi v_0^2 E / c^3$$

where c is the velocity of light and E is energy of the average oscillator. Notice that this crisp expression doesn't involve any information about the particular device Planck used to simulate atoms (the electron charge or mass, for example). Only the natural frequency appears. Because of its extreme generality this result must have encouraged Planck that he was on the right track.

Next, Planck had to calculate E, the energy of the average oscillator. He followed much the same statistical mechanical procedure described earlier in connection with the Maxwellian distribution. In effect he searched for the most probable distribution of a fixed amount of energy *total* among a fixed number of oscillators N. This would lead to the desired black body radiation spectrum.

At the very last step he ran into a problem. In order to predict the spectrum experimenters had found, he had to admit that an oscillator can only emit or absorb light in finite bits ("quanta") whose elementary energy equals $h\nu$, where h is Planck's universal constant. This ad hoc assumption conflicted with everything he knew. He spent many years trying to obtain the same result using only classical physics. Although he failed in this effort, he triumphed in revealing the quantum nature of light.

Planck's final formula for the energy density of radiation with frequency ν is

$$U(\nu) = 8\,\pi\,h\,\nu^3 \,/\, c^3 > [\, e^{\,h\nu/kT} - 1 \,]^{-1}$$

and is illustrated in Figure 3.1.

In 1924 Albert Einstein was able to derive this spectrum of black body radiation more rigorously, using a new result of Satyendra N. Bose on the correct statistics of quantized photons.

3.6 The Ionized States of Heavy Atoms

Recall that the sun contains a mixture of elements (Table 1) in which hydrogen and helium are most abundant and in which the other heavier elements appear only as traces. Each heavy ion consists of a nucleus of protons and neutrons, surrounded by shells of bound electrons. The inner shells are tied most closely to the nucleus, the outer shells less so.

When the interior of the sun reached its present high temperature, collisions among the atoms became so energetic and so frequent that atoms "ionized"— they lost all or many of the outer shells of bound electrons. The electrons, being light and energetic, were free to wander everywhere.

At temperatures in the millions of kelvin, hydrogen and helium have lost all their electrons, and exist as protons and alpha particles. The heavier atoms retain some of their more tightly bound electrons, however. Iron, for example, with an original complement of 57 electrons, now exists in several forms, with 37, 38, 39, or 40 bound electrons. The problem astrophysicists faced in the early 1900s was how to estimate, for each type of atom, the fraction that have lost a given number of its original electrons. In the interior of a star, where thermodynamic equilibrium prevails, the problem becomes much simpler. There one could be sure that *the most probable situation* would prevail. One didn't need to know the messy details of *how* equilibrium would be established—the specific interactions and atomic cross-sections. Instead, one could apply Boltzmann's procedure of distributing N

electrons among M energy shells and look for the most probable arrangement. This task was carried out by Mahindra N. Saha and Ralph H. Fowler in the early 1920s. They predicted general formulas for the ionization equilibrium of a generic atom.

Figure 3.8 shows a sample of their results for iron. At a fixed electron density and as the temperature increases, iron atoms lose more and more bound electrons. At any fixed temperature, as many as three different iron nuclei can exist, however. For example, at a temperature of 10 million kelvin, iron nuclei which have lost 17, 18, 19, or 20 electrons all coexist in different proportions. Heavy ions with their bound electrons can contribute substantially to the opacity of the plasma to X-rays. A photon with enough energy can liberate a bound electron and send it flying free from its parent nucleus. In this process the original energy of the photon is completely expended in tearing loose the electron and in giving it a final velocity. The photon has been absorbed in a so-called "bound-free" process.

This cannot proceed indefinitely without all the ions being totally stripped. The inverse process of electron capture proceeds at just the necessary rate to establish an equilibrium. In this inverse *emission* process, a photon is created from the original kinetic energy of a fast electron.

Chapter 5

Looking Inside from the Outside

5.1 Refraction of Sound Waves

Figure 5.3 illustrates how a wave is refracted. This wave has just been reflected from the surface and is traveling downward at an angle. Because the plasma temperature increases inward, so does the speed of sound. The part of the wave nearer the surface travels in relatively cool plasma, where the speed of sound is relatively low, while the deeper part travels in hotter plasma where the speed is greater. So the deeper part of the wave moves faster and the wave pivots, like a line of soldiers turning a corner, and heads back toward the surface.

5.2 Standing Waves

Figure 5.5 doesn't show a sound wave, it shows instead a standing wave on a violin string. The string is free to vibrate up and down between fixed positions, the "nodes." This standing wave can be thought of as two waves that travel in opposite directions and whose sum is the standing wave. In Figure 5.5 we see that as the twin waves separate, they always overlap in such a way that the nodes remain stationary. This is the essence of a standing wave—the nodes remain stationary in space, while the medium between them (the string of a violin or the plasma of the sun) oscillates.

A sound wave in the sun is free to travel in any direction: in or out, north or south, northwest by west, and so forth. But only those waves with particular wavelengths and directions can form standing waves and survive random collisions with other waves. To form a standing wave pattern, or "mode," the wavelength of each twin wave must fit into the circumference of the sun a whole number of times (e.g., 157) and also fit into the depth of its spherical shell a whole number of times (e.g., 24). Only then will the twins overlap in three dimensions so as to produce fixed nodes.

Each one of the standing wave patterns (or "mode") can be given a label that consists of its frequency and the number of nodes it possesses in three different directions. The number (n) of nodes in the *radial* direction is called the order, the number (l) of nodes along a *meridian* is called the degree, and the number (m) it has *along the equator* is called the azimuthal number. Figure 5.7 shows a number of patterns that correspond to different values of l and m.

5.3 Modes and Nodes

All the standing waves on a single curve (modes) have the same number of nodes (stationary points) along a solar radius. Some waves have as many as 40 of these radial nodes. Next, each curve is a string of individual wave patterns or "modes" (they are resolved as little dots in the lower left of Figure 5.2). Each of these modes has a different number of nodes (called "degrees") along the *meridian* of the sun. Waves with as many as 500 meridional nodes have been observed.

Finally, each wave can be classified by the number of nodes it has along the solar *equator*. Waves with 1,000 of these nodes have been seen.

So in summary, any particular wave can be labeled with the numbers of nodes it has along the solar radius, meridian, and equator. There at least $40 \times 500 \times 1,000$ or *20 million* different modes vibrating in the sun at any moment!

5.4 Instruments Used in Helioseismology

Several types of instruments can be used to measure plasma velocities. In general they all isolate narrow spectral lines and allow astronomers to measure their Doppler shifts. *Spectrographs* use so-called diffraction gratings to disperse sunlight into its component colors and to display the narrow absorption lines that riddle the spectrum. A spectrograph displays a complete spectral line at each point along a line on the solar surface. In contrast, an optical *filter* displays a two-dimensional image of the sun at a single wavelength within a spectral line. Another type of filter, *the resonance absorption cell*, uses the vapor of an element like sodium or potassium to isolate a spectrum line. It accepts light from all parts of the solar disk simultaneously, without resolving any surface detail. In effect, it treats the sun as an unresolved star.

These instruments must have superb resolution in wavelength to be useful. A plasma velocity of 100 m/s produces a wavelength shift of only three parts in ten million, or about .0002 picometers. Spectrographs and optical filters can detect a meter per second but will drift by more than that in a few hours. Resonance cells are much more stable and can detect speeds as low as a few millimeters per second, but can only observe in a single spectral line.

5.5 How to Find the Internal Rotation of the Sun

We posed the question of just how the internal rotation of the sun has been determined and what is the physics that lies behind it. The answer lies buried in the diagnostic diagram, Figure 5.2. Individual modes, specified by their quantum numbers n and l, appear as small dots in the left side of the diagram. If we magnify one of these (Figure 5.13) we see that it is really a symmetrical cluster of smaller dots, each at a slightly different frequency. Each one of these smaller dots corresponds to a different azimuthal mode number m. The basic mode frequency has been "split" by the rotation of the layer in which the mode is trapped. *The frequency spacing between the smaller dots is a direct measure of the rotation frequency of the layer.*

To understand why these additional frequencies appear in the oscillation spectrum, we need to review a little. Recall that each mode of vibration of the sun corresponds to a three-dimensional standing wave pattern of sound waves that are trapped in a spherical shell. At the surface the mode appears as a *two-dimensional* standing wave pattern. Figure 5.7 displays several of these patterns, among the millions of patterns that appear at the same time and overlap at the surface. The quantum numbers l and m specify the number of nodes of the pattern on a meridian and on the equator, respectively.

So far, so good. But *why* do the mode frequencies split? The reason is that sound waves propagating in a moving medium are carried with the medium. This effect is called "advection." You will remember that a standing wave consists of twin traveling waves, that travel in a series of loops, in opposite directions around their shell (Figure 5.5). The twin that travels in the *same* direction as the rotation (the "prograde" wave) will be carried along by the plasma, like a boat moving downstream with the current. The prograde wave will therefore sweep westward *faster*, relative to an observer on earth, than it would if the shell didn't rotate. Hence, the observer sees the twin's frequency shifting upward slightly. The other twin (the "retrograde" wave) is in effect moving upstream *against* the direction of rotation. It will travel more slowly than it would if the sun didn't rotate. Hence, the observer sees this twin's frequency shifting downward slightly. The single mode frequency has been split into two. The net result is that the twin waves still overlap nicely to form a standing wave pattern but this pattern *rotates* with the shell.

The advance or retardation of the twin waves also shows up as small shifts in the basic oscillation frequency. Let's see how these extra frequencies appear.

In principle, we could measure the frequency of either the prograde or the retrograde traveling wave of a mode by counting the number of its nodes that pass a fixed north-south line on the solar disk in a fixed amount of time (Figure 5.14). If the sun were *not* rotating, two nodes of a traveling wave would pass this line about every five minutes, because that is the basic period of the oscillation. We would see a stationary wave pattern, like the one in Figure 5.8 with $l = 7$ and $m = 7$, that oscillates at a frequency around 3 millihertz.

But the sun's interior *does* rotate. Suppose the shell containing the $l = 7$, $m = 7$ mode rotates once in 27 days or at a frequency of 0.4 microhertz. As the shell rotates from east to west, the prograde wave is carried westward, and therefore we would count 7 *additional nodes* in a rotation period of about 27 days. These extra nodes would show up in a Fourier analysis of the oscillations as a small increase of frequency, namely, 0.4×7 or 2.8 microhertz. Similarly the retrograde wave would *lose* 7 nodes in 27 days and this would produce a corresponding decrease of frequency. The net result is that two additional frequencies appear in the oscillation spectrum, one at 3 millihertz *plus* 2.8 microhertz and the other at 3 millihertz *minus* 2.8 microhertz. The basic oscillation frequency has been *split*.

Moreover, the $m = 1, 2, 3, 4, 5$, and 6 modes would also appear as pairs of shifted frequencies, equal to $\pm m \times 0.4$ microhertz. These pairs form the cluster we saw in Figure 5.13. By measuring the frequency shifts within this cluster, a helioseismologist can derive the rotation speed of the shell that contains the mode.

Of course, he or she must rely on a detailed theory of solar oscillations to identify the mode and the depth of the shell in which it lies. But it is this exquisite blend of theory and observations that yield the marvelous map of Figure 5.12.

Chapter 6
THE HOT ATMOSPHERE

6.1 How Edlèn Did It

First let's look at some heavy ions. Figure 6.5 shows the energy levels of three related ions. Lithium (labeled Li I) has its usual set of three electrons; beryllium normally has four, but in this case has lost one to become the ion "Be II"; and carbon normally has six, but in this case has lost three to become "C IV." All three ions are left with the same number of electrons, three. They form a so-called "iso-electronic sequence."

Notice how similar the patterns of energy levels are. And notice how the energies of the corresponding levels increase as we move up the sequence. The second level, for example, changes from 1.85 ev in lithium to 3.96 ev in beryllium II to 8.01 ev in carbon IV or by about a factor of two for each step up the sequence. And the corresponding spectral lines shift systematically into the far ultraviolet range.

The next ion in this sequence would be a nitrogen atom that has lost four of its seven electrons, $N\ V$. Now that we know how the energy of the second level increases as we move up the sequence, we could easily extrapolate to find the energy of the second level in the ion nitrogen V.

This is the bootstrap process Edlèn used to find the energy levels of very highly ionized atoms, like iron that has lost 14 or 15 electrons. He would produce ions within the same iso-electronic sequence, one at a time, measure their spectra, and extrapolate to find the energy levels of the next ion, using the regularities within the sequence to interpret his results.

6.2 Estimating Plasma Density in the Transition Region

In any sample of plasma only a small fraction of the mass (approximately 1 part in 100,000) consists of carbon. Where the temperature reaches 100,000 kelvin, a large (and calculable) fraction of the carbon is in the form of ions with three electrons missing, C IV. Of those, only a calculable few are capable of emitting specific spectrum lines, say, the ones at 155 nm. So if one measures the amount of light received in those lines at that temperature, one can work backward to learn how many carbon ions are emitting along the line of sight through the atmosphere, and this is directly related to the total amount of plasma. If one then takes into account the length of the line of sight, one can derive the average density of the plasma. There are other ways, of course. For example, C IV emits two lines at 154.8 and 155.0 nm and their intensity ratio depends on the local plasma density. So if you can measure both lines simultaneously, you have an independent way of estimating density.

6.3 An Alternate Approach to Energy Losses

Athay and Anderson didn't try to reproduce the complete solar spectrum, as the Harvard team did. Instead, they searched for a model atmosphere in perfect energy balance that has the same temperature distribution as the Harvard model (Figure 6.9). So they varied the deposition of nonradiative energy in model solar atmospheres, and calculated the resultant temperature distributions solely by balancing energy gains and radiative losses. When they found a model temperature distribution that matched the Harvard model, they were then able to determine the required distribution of nonthermal energy. Their total energy requirement is about twice that of Avrett and company, and is lost primarily by thousands of weak spectrum lines of iron.

6.4 Heating the Low Chromosphere

Several groups have attempted to determine the flow of acoustic power through the low atmosphere, say, below a height of 2,000 km. The usual technique is to measure the periodic Doppler shifts in a pair of spectra lines that sample different heights in the atmosphere. Then they try to determine how much of the observed oscillations are correlated, so as to represent a traveling wave. With a little arithmetic, they can then estimate the acoustic power at different heights.

This is a very uncertain business, however. Different observers arrive at very different estimates for the amount of acoustic power flowing up from the photosphere. However, they do agree that above a height of about 2,000 km the remaining sonic flux is too small to match the losses of the overlying atmosphere.

Athay's and White's conclusion has been confirmed: another source of heating seems to be required above the mid-chromosphere.

Chapter 7
THE MAGNETIC ATMOSPHERE

7.1 Thousand-Gauss Magnetic Fields

In 1973, Jan Olaf Stenflo, a Swedish solar astronomer, proved that the supergranule network far away from active regions also contains kilogauss fields. He made simultaneous measurements of the photospheric field in a small area, using two spectrum lines, one more sensitive to the field strength than the other. The two lines yielded *different* estimates of the *same* field strength. A number of explanations are possible for this result but not if the two lines are chosen from the same family of lines. In that case only one explanation was possible: the observed area is not uniformly filled with magnetic flux. From the two measurements, Stenflo could find both the field strength and the total flux. He estimated that the strength was about a thousand gauss and that all the flux occupied less than 10% of the observed area. The real flux elements could be as small as a few hundred kilometers across. Perhaps elementary flux tubes do exist!

Surprisingly, these results were very similar to those found in active regions. The only difference is that an active region has more magnetic flux, that is, more flux elements more closely packed.

7.2 Biermann's Scheme for Kilogauss Fields

When a fat bundle of magnetic flux erupts through the photosphere, its magnetic pressure resists horizontal convective motions of granules in its neighborhood. As a result, the granules cannot deliver heat to the bundle. Neither radiation nor heat conduction from the granules is sufficient to replace the radiative losses of the bundle. So the bundle cools off and contracts. The contraction concentrates the flux, the field strength rises, and a sunspot forms with 3,000-gauss fields.

7.3 Parker's Scheme for Kilogauss Fields

This is the way Parker's mechanism works. Imagine that the flux tube settles into a vertical position just after it emerges from below and that the plasma inside it is in hydrostatic equilibrium. Next, the convective supply of heat is cut off. As the column of plasma inside the tube cools, it must readjust its vertical density distribution to *try* to come to a new equilibrium. The plasma must flow downward, to match the pressure of the external gas deeper in the photosphere. But since the plasma in the tube receives very little heat from outside the tube, its temperature will always remain lower (and its density higher) than that of its surroundings, no matter how deep it sinks, and therefore it will *continue* to sink.

As a result the upper part of the tube will be evacuated. In this upper part, the magnetic pressure must increase to compensate for the loss of plasma pressure and so the flux tube will pinch in. The diameter of the flux tube will shrink to perhaps 100 km at the top of the photosphere, and the field strength there will rise to a thousand gauss or more.

7.4 Spectroscopic Polarimetry of Flux Tubes

Their method is to measure the wavelength variation of polarization inside two or more magnetically sensitive spectrum lines, at various positions across the solar disk. They strive to obtain highly precise line profiles, and very sensitive polarization measurements, even at the sacrifice of some spatial resolution. For most of their observations they have used the Fourier Transform Spectrometer at the Kitt Peak National Observatory, a special instrument capable of producing exquisitely detailed spectra.

To interpret their observations they adopt a two-dimensional model flux tube that looks either like a sheet or like a cylinder of magnetic flux (Figure 7.8). The tube is surrounded by the field-free plasma of the photosphere. Then they adopt a trial temperature distribution inside the tube, assume the plasma is in hydrostatic equilibrium, and calculate the detailed polarized spectrum lines they have observed. If the predictions and observations disagree, they can modify the size of the tube, the amount of flux it contains, and the temperature distribution inside it, until they find agreement. They can add some extra details, like the turbulent velocities inside the tube, to achieve a good match.

They have gradually added more and more observations to further constrain their models. First they observed only two spectrum lines, then as many as twelve, including several very sensitive lines in the near infrared spectrum. They first observed only the center of the solar disk, then regions between the center and the limb. As the data accumulated, and more subtle details appeared, their models had to be improved.

7.5 Birth and Death on the Magnetic Network

Schrijver and company measured the distribution of fluxes of magnetic patches in the network. The distribution follows an exponential law such that patches with a flux of 1×10^{18} maxwells are a hundred times more numerous than patches with 5×10^{18} maxwells. Then they tried to account for the distribution with a numerical model. The model assumes that all the flux arriving in the network originates in the ephemeral active regions (EARs) whose rate of emergence has been determined independently. The arriving flux merges with flux of the same polarity or cancels with flux of opposite polarity. By adjusting the rate of cancellation (which is less well known than the emergence rate) they were able to reproduce the observed size distribution of fluxes.

Chapter 8
THE MIDDLE KINGDOM

8.1 Hollweg's Spicule Model

Hollweg and colleagues started with a vertical flux tube whose internal plasma was in hydrostatic equilibrium. Then they gave the bottom of the tube two gentle twists, 2 km/s in transverse speed and 90 seconds long. The twists generated two Alfvèn waves, which traveled up the tube. As the waves encountered lower plasma densities, their twisting motions accelerated until their speeds reached sound velocity. Then the waves steepened and shocked. A whole train of complicated hydrodynamic shocks were then triggered. Some of these propagated into the corona and according to Hollweg, could heat it.

As the waves rode up the tube, their magnetic pressure pushed the plasma ahead like a snowplow. A plug of plasma reached a speed of 18 km/s and its top reached a height of 6,500 km after 800 seconds. Then it started to fall back. These properties, as well as the predicted plasma density, resembled real spicules. In fact, any twist exceeding a kilometer per second in a field strength of about 10 gauss and lasting less than two minutes would produce a recognizable spicule. In this respect the model supports the idea that Alfvèn waves produce spicules.

8.2 Haerendel's Spicule Model

An Alfvèn wave riding up a flux tube accelerates the electrically charged ions. In a fully ionized gas, a plasma, all the ion kinetic energy is returned to the wave in one cycle; the wave loses no net energy. But in a *partially* ionized gas, there are neutral atoms present, which the wave cannot accelerate. However, the accelerated ions collide with the neutrals, dragging them along. But there is some slippage between the ions and neutrals because there is a finite time interval between collisions. Because of this slippage, two related things happen. First, the wave loses energy to the plasma, and heats it. Secondly, the wave loses momentum. In effect it pushes the plasma up the tube with its wave pressure.

Haerendel proposed that such pressure, along the tube direction, could support a column of partially ionized gas against the pull of gravity. A little stronger pressure could gently accelerate the gas to spicule speeds. The beauty of his proposal is that the pressure is applied continuously and all along the tube.

8.3 Carbon Monoxide

Ayres and company had used a special form of spectrograph (the Fourier Transform Spectrometer) at the Kitt Peak National Observatory to map the CO bands at 4.7 and 2.3 microns, simultaneously with the profile of the calcium K line, in a patch of the quiet sun. Recall that the K line is the classic probe of the chromosphere and was used extensively in constructing the conventional temperature maps.

Carbon monoxide (CO) is only one of many simple molecules that reside in the solar atmosphere. Others include H_2O (water), CH, C_2, and OH. Such molecules can only exist where the temperature is cool enough. Carbon monoxide, for example, splits into its constituent atoms of carbon and oxygen at temperatures above 5,000 kelvin. Moreover, the CO molecule couples strongly to the local electron temperature and so its energy levels and spectrum behave as in thermodynamic equilibrium.

The spectra of CO are "bands" of hundreds of sharp absorption lines in the infrared end of the solar spectrum (Figure 8.3). The deepest lines lie in a band around 4.6 microns, well beyond the limit of visible light. As Milne and Eddington taught us (recall their argument from chapter 6), the brightness in the cores of such lines is a direct measure (the Planck function) of the gas temperature, at the height where the core photons escape. So the line cores are good thermometers.

8.4 Athay's Observations

The spectral lines of these atoms switch from absorption (on the disk) to emission at the limb. The details of the transition are sensitive to the chromospheric temperature profile and to the fractional area.

8.5 Transition Region Explosions

Every 40 seconds on average, an explosion occurs in the TR somewhere on the sun. A jet of plasma erupts at speeds ranging from 100 to 400 km/s, with plasma temperatures that run the gamut from 20,000 to 200,000 kelvin. These explosive events occur preferentially in strong gradients of magnetic field, at the borders of supergranules. We can speculate that their energy is derived from the partial annihilation of some of the flux. Indeed, the NRL group tried unsuccessfully to identify these events with the source of heating for the corona.

Chapter 9

ACTIVE REGIONS

9.1 Evolution of an Active Region

About this time or a little later, bright "faculae" appear in white light over the legs of the loops, indicating enhanced heating by some means. Within another day the loops become hot enough to emit extreme ultraviolet light (30–120 nm) and soft X-rays (0. 2–6 nm). They begin to resemble their portrait in Figure 7.1.

Most emerging regions stop growing in a day and remain as simple bipoles, containing less than 10^{20} maxwells. But in a few, new bipoles continue to bubble up near the center of the region. These new bipoles merge or link with the older bipoles, to form multipolar activity "complexes" that contain as much as 2×10^{22} maxwells. Such complexes can live for many months, slowly decaying and then being rejuvenated by fresh deliveries of flux.

Strong horizontal flows appear in the photosphere as a region breaks through the surface. In a well-studied case, a powerful upwelling of plasma accompanied the emergence. Photospheric plasma spread out from the upwelling, dragging pores and smaller flux elements with it. The pores then streamed along the edges of the region toward sunspots of the same polarity, while the spots moved apart. The forces that drive such eccentric motions are complicated and not well understood. The upwelling is probably driven by buoyancy, but with such strong fields (several hundred gauss) magnetic forces in the evolving rope may also play a role.

9.2 Active Region Temperatures

To determine how all this plasma is arranged in an active region, astronomers make images in different spectral lines, chosen to represent different plasma temperatures. The hot plasma is only detectable in the extreme ultraviolet or X-ray bands so that images must be made with equipment aboard rockets or satellites. Real progress with this technique began with Skylab. Several groups have been active recently, combining data from two or more instruments aboard YOHKOH, SOHO, SERTS, and TRACE.

Figure 9.5 shows an active region imaged from SOHO in several lines that span the range of 20,000 to 2 million kelvin. Several characteristics stand out. Few loops are entirely "cool," less than 1 million kelvin, or entirely "hot," greater than 1 million kelvin. Instead, most loops have both hot and cool plasma. The cool plasma resides in the legs, the hotter material at the tops.

YOHKOH has revealed the properties of the hottest active region plasma, between 3 and 10 million kelvin. Here too, the "cooler" plasma (around 3 million kelvin) is found in the legs of loops, while the tops contain 4 to 5 million plasma. Contrary to intuition, the brightest loops may be cool, the faintest hot. This may imply that the cool loops are denser than the hot ones. And, most interesting, the hottest loops—5 to 8 million Kelvin—are also the ones with the shortest lives, usually less than three hours.

9.3 The First Scaling Law for Coronal Loops

These few principles, plus the assumption of constant plasma pressure throughout the loop, were sufficient to derive a relation between the length L, the pressure p and the maximum temperature T_m in a loop.

$$T_m = C(p\,L)^{1/3}.$$

This simple relation fitted the data from Skylab reasonably well, and with a change in exponent from 1/3 to 1/5, it fits the recent data from the YOHKOH satellite. The constant C depends on the details of the heating mechanism.

9.4 Force-Free Magnetic Fields

The tangled fields you see in Figure 9.6 are *force-free*. A few words of explanation are in order. A magnetic field, in which *no* electric current flows (a "potential field") is smooth, like the "iron-filings" field of a bar magnet. In contrast, the sheared and twisted shapes of chromospheric loops convinced astronomers long ago that electric currents must flow in the chromosphere and into the corona. When detailed vector magnetic maps of the photosphere became available, they also showed twisted field lines that pointed to the presence of electric currents.

If an electric current flows *across* a magnetic field, a "Lorenz force" develops which will tend to distort the shape of the field unless some other force counteracts it. But in the corona the plasma pressure is far too low to resist such Lorenz forces. The only way the coronal fields can be "force-free" is if the electric current flows *along* the direction of the field, along the field lines.

Mikic and his colleagues assumed this condition prevails in the coronal field of active regions. From observations of the photospheric vector field they were able to determine the distribution of electric currents and use these to compute the fields shown in Figure 9.6. Field strengths of a few hundred gauss were calculated at a height of about 10,000 km above the photosphere.

9.5 Sunspot Models

Such models were built over a decade ago. Figure 9.8 shows one example, by Vic Pizzo, then at the High Altitude Observatory. He used observational models of density and temperature in the low-spot atmosphere to constrain a force-free magnetic field. The model reproduces several important features of a real spot, including the so-called "Wilson Depression" in the umbra. Because the plasma is cool there, one sees down to a deeper depth than elsewhere, and the umbra looks like a small pit in the photosphere. (The heavy line in the figure shows the level to which one sees.) Throughout most of the umbra the field is nearly current-free, but not in the penumbra, where it is closer to force-free.

Waves and Oscillations in Sunspots

The photospheric waves in the umbra are so-called "fast modes," in which the restoring force is a combination of magnetic and plasma pressure. They travel at a compromise speed between the speed of sound and the Alfvèn velocity, about 10 km/s. They are trapped, in a kind of cavity (see Figure 9.11) that is determined by the way the sound speed and Alfvèn speed vary with depth and height in the umbra.

The chromosphere of an umbra also oscillates, but at periods near three minutes, not five. As Figure 9.11 shows, these oscillations also represent waves that are trapped in a second, higher cavity. The connection between the two cavities and the way in which these chromospheric oscillations are excited, is uncertain at the present time.

Two types of conceptual models have been proposed to explain penumbral waves. In one type, the waves are fast-mode waves, that travel in the lower cavity of the spot. In the other type of model, the waves are thought to be *surface* waves on the sharp subsurface boundary between the strong fields of the penumbra and the surrounding field-free photosphere. Each model has its good points, but neither takes account of the recent observations of the filamentary and interleaved structure of the penumbral field.

Chapter 10
EXPLOSIONS ON THE SUN

10.1 How Fast Reconnection Works

Figure 10.9 shows a simplified arrangement, originally proposed by H. E. Petschek in 1964, that shows how this could happen. In this picture, all the spaces between oppositely directed field lines are filled with plasma. Suppose that the field lines, together with their entrained plasma, push toward each other from the top and bottom of the figure. The plasma at the interface between the opposed fields is squeezed sideways, out of the intervening space. When the distance separating the opposed fields is small enough (no longer than a football field) the fields can diffuse across it rapidly and reconnect.

Now the reconnected field lines (the hairpin loops at each side) can escape sideways. As they escape, their tension propels their entrained plasma sideways in two fast jets, like a slingshot. Eventually, the jets dissipate their kinetic energy as heat.

At the same time, an induced electric current flows perpendicular to the page in the reconnection zone, and the electrical resistance of the plasma there converts the current to heat. The plasma in this tiny zone will heat up impulsively to huge temperatures and we have a flare.

10.2 Anomalous Resistivity

In normal plasma, an electric current consists of slowly drifting electrons. Imagine that the electrons are like cars on a freeway. The volume of traffic depends on the number of cars per mile and their speed. Similarly the size of an electrical current depends on the number of available electrons and their drift speed. The faster the electrons (or cars) move, on average, the larger is the current (or volume of traffic) they can carry. When the electrons collide with slow ions in the plasma (accidents with trucks!), they lose their momentum and have to start over again from rest. That is the origin of the plasma's "normal" electrical resistance, and also the origin of traffic slowdowns.

In the reconnection process, very large currents are suddenly induced in a very small volume. (To continue our analogy, the cars must now pass through a bottleneck.) Since the volume can contain a limited number of electrons at any one time, the current they carry can only increase if they accelerate.

But if the electron drift speeds reach the velocity of sound in the plasma, a natural threshold, the whole situation changes. At that point, so-called ion-acoustic waves are induced in the plasma, and these waves act as microscopic barriers to the drifting electrons. (Similarly, in our analogy, as cars ahead of you slow down to pass through the bottleneck, a *wave of slowing* reaches you and you must also slow down.) As a result, the resistance of the plasma increases enormously and the rate of dissipation of current accordingly increases. With this new factor of anomalous resistivity, theorists think they can explain how flares heat up so quickly.

10.3 A Hybrid Flare

Soft X-ray images of the loop showed that it originally had a cusp-like extension. Following Sturrock's ideas on two-ribbon flares, Masuda interpreted that cusp as a current sheet, where oppositely directed magnetic field lines could reconnect. To explain his observations, he proposed the magnetic geometry shown in Figure 10.11. Reconnection in the cusp (well above the loop) supposedly produced supersonic jets (with speeds of 3,000 km/s) that rocketed downward and collided with the top of the closed loop. A shock wave formed there and the plasma was heated to an incredible temperature (200 million kelvin!). Hard X-rays (53 kev) were emitted from this super-hot region as *thermal* radiation.

To explain the appearance of bright patches of soft X-rays, at the feet of the loop, Matsuda invoked fast electron beams. These beams were ejected from the reconnection site above the loop, streamed the legs of the loop, and bombarded the chromosphere. So, according to Matsuda, this flare was a hybrid that emitted both thermal and nonthermal X-rays.

Chapter 11
THE CORONA AND THE WIND

11.1 A Possible Cause of a Streamer Eruption

According to Low, the cavity is buoyant and tends to rise spontaneously, like a balloon, because it is less dense than the surrounding streamer. It is anchored in two ways, however. The weight of the prominence, hanging in the cavity's helix, acts as ballast for this balloon. And the row of magnetic arches that enclose the cavity act as tethers to hold it down.

In time, these two restraints can weaken. As supergranules shuffle the feet of the arches, spreading them apart, the arches are forced to expand and the magnetic tension they exert on the cavity decreases accordingly. At the same time, the prominence may leak some of its mass back to the chromosphere, reducing its ballast effect. The combined effect is to allow the helix to expand and rise slowly. Eventually the arches rupture and the helix is free to blow into space. The prominence is dragged along in the coils of the helix.

Art Hundhausen, a colleague of Low's, has actually observed this process in sequences of white light images. The lower parts of streamers swell over a few days and reach a critical size before they erupt.

In Low's scenario, the energy to propel the CME is contained in the twist of the magnetic helix and in its buoyancy, not as the heat of an explosion. His scheme also accounts for the slow acceleration of the CME. Low agrees with Rust's idea that CMEs carry off most of the magnetic flux and helicity that the sun creates during its eleven-year activity cycle, but he extends the picture to include the reversal of polarity of the solar poles.

11.2 How to Find Alfvèn Waves?

One method depends on the width of coronal spectral lines, usually measured in nanometers. The line width one observes depends on the motion of the ion that emits it. The faster the ion moves along the line of sight, the wider is the line. In hot plasma, the thermal speed of the ion accounts for much of the observed line width. For example, the average thermal speed of an iron ion in a 2 million kelvin corona is 31 km/s, so that a line at 28.4 nanometers should have a width of about 0.006 nm. But this line is typically broader.

To account for the difference, astronomers commonly invoke the presence of Alfvèn waves, with oscillatory motions of 20 to 40 km/s. But other explanations, such as purely turbulent motions, are possible, so that the existence of Alfvèn waves is still not definitely proven.

Suppose they do exist, however. Then another problem with them arises—they are difficult to turn into heat. They don't compress the plasma, so shocks don't form. And although many other ways of dumping Alfvèn wave energy have been proposed, only one ("resonant absorption") has gained any credibility. Unfortunately, it only works with waves that have short periods (10–100 seconds) and that raises the third problem: how to generate Alfvèn waves with the required periods and with the required strength.

11.3 Nanoflares Again

In a very careful recent study, a group of astronomers repeated Shimizu's count of nanoflares, using extreme ultraviolet images from TRACE. They too concluded that the number of observed nanoflares with energy as small as 10^{12} kilowatt-hours was too small by a factor of about 300 to heat the quiet corona.

11.4 The Source of the Wind's Energy?

Don Hassler (from the Southwest Research Institute in Boulder, Colorado) and his international team used SUMER, an ultraviolet spectrograph aboard the SOHO satellite, to measure the outflow of coronal plasma on the disk of the sun. They detected steady outflows of five to six kilometers per second in a spectral line of neon 8 (Ne VIII), a coronal ion that forms at a temperature of 600,000 kelvin. But even more important, they found the flows are *restricted to the borders of supergranules*, just as Parker had suggested. Evidently the annihilation of small-scale magnetic fields at the supergranule borders can release enough energy to start the wind.

And as you may have noticed, this same mechanism has been invoked (in the form of the "magnetic carpet" scenario) to heat the quiet corona.

Chapter 12
"LIKE A TEA TRAY IN THE SKY"

12.1 The Hanle Effect

The hydrogen and helium atoms in a prominence absorb and then re-emit light from the underlying chromosphere. Because the prominence is illuminated only from below, part of the light it emits sideways is *polarized*. This means that the electric vibrations of the light are confined to a single plane in space, instead of being distributed randomly in all directions. (Imagine a child snapping one end of a rope up and down. The waves that travel along the rope are then polarized to lie in a vertical plane. If she snaps the rope end in random directions, the waves are then unpolarized.)

Now, the magnetic field in a prominence tends to reduce the amount of polarized light that is emitted. This is the Hanle effect and it arises because ions in the prominence plasma can spiral around the field lines several times before emitting the photons they have absorbed. In effect the field randomizes the planes of polarization.

12.2 How to Make Sinistral and Dextral Filaments

They imagine that differential rotation below the photosphere (the heavy arrows in Figure 12.4a) pulls the large-scale field into an inclined pattern, as shown, while the coronal field crosses the (dotted) polarity inversion line (PIL) in the north-south direction. In the next step (Figure 12.4b), small magnetic loops rise buoyantly to the surface from below. Their roots have polarities opposite to that of the large areas to the left and right of the PIL.

Then the loops link up to form a channel by reconnecting at their roots (Figure 12.4c), a process that has actually been observed by Sara Martin. The channel has the correct orientation with respect to the PIL to form a dextral filament in the northern hemisphere. Finally a dip in the channel fills somehow with plasma to create a visible filament. The panels on the right side of the figure demonstrate that the scheme works equally well for sinistral filaments in the southern hemisphere.

Chapter 13

SOLAR AND STELLAR CYCLES

13.1 How the Poles Reverse Polarity

Because the magnetic ropes slope in latitude, all the follower spots emerge slightly poleward of their leader spots (as in Figure 13.1), with the *opposite* magnetic polarity of the nearest solar pole. Therefore, as the followers diffuse poleward (presumably through the shuffling action of supergranules), a small fraction of follower flux (about 1%) overwhelms the existing polar polarity. The poles then reverse their polarities, just as observations require. The time required for the new flux to diffuse to the poles determines the period of the cycle. (Incidentally the two poles may not reverse simultaneously. For short periods of time, the sun may have two north magnetic poles!)

When the polar fields reverse, a new poloidal field develops, stretching from the north to the south poles. In Babcock's model, it must submerge into the convection zone somehow to start the next cycle.

13.2 The Coriolis Force

The Coriolis force is an apparent force that an observer in any rotating system experiences. Imagine a storm system resting on the earth's equator. Because the system participates in the earth's rotation, it has a speed toward the east of about 1,000 miles per hour, relative to an observer fixed outside the earth. An observer on the earth at mid-latitude has a smaller easterly speed because of her rotation and one at the pole has zero easterly speed.

Now suppose the storm starts to move north. And suppose it keeps its initial easterly speed of 1,000 miles per hour. Then when it arrives at mid-latitude it will be moving east faster than the observer sitting there. Therefore she will see the storm swing eastward as though a force were pushing it east. That apparent force is the Coriolis force. It deflects moving objects to the east in the northern hemisphere and to the west in the southern hemisphere. It is the cause of cyclonic storm systems.

13.3 Dynamo Waves

Imagine that early in the cycle (Figure 13.4a) a toroidal rope exists at high latitude in each hemisphere. The alpha effect twists loops along the rope, giving them a polar component (the little circles) (*b*). The loops combine, as in *c* to form circular fields that enclose the original rope.

Now suppose that the deeper layers of the convection zone rotate *faster* than the higher layers. In other words, suppose the angular speed of rotation decreases outward along a solar radius. If that were true, the circular field lines would be *tilted* as in *d*. They reconnect to form two toroidal ropes with opposite polarities, on either side of the original rope (*e*).

Next, the northernmost loop could diffuse into the original rope and annihilate it, leaving a small remainder (f). Notice that the original rope would be replaced by one with the same polarity but at a lower latitude. In effect, the toroidal field would migrate toward the equator!

As this process repeats, the rope remaining after each repetition approaches ever nearer to the equator. Finally the ropes in the northern and southern hemispheres cancel, leaving a rope at high latitude with a polarity opposite to the original rope (compare Figures 13.4a and g).

13.4 Gilman's Computer Model

As a first step, he designed a numerical model for the internal differential rotation of the sun. He assumed the convection zone contains "giant cells" that extend the full depth of the zone and that drive a chain of smaller cells. This assembly of cells carries the heat toward the surface, where it is radiated away. Coriolis forces and turbulent diffusion affect the giant cells in such a way as to form long convective rolls (called "banana cells") that extend from pole to pole. The net effect of convection is to redistribute angular momentum throughout the zone, and therefore create differential rotation.

13.5 The Effect of the Dynamo Number

To explore these changes, let's imagine that we have a toy star, and that we can crank up its dynamo number, by spinning the star faster and faster. If the star starts from rest, there is no dynamo action, no magnetic field generation until the rotation rate and D reach a critical value $D1$ (see Figure 13.9, top). As D is increased further to Dz (Figure 13.9, middle) the star develops clean cycles. Finally, when we reach the next critical dynamo number, $D3$, the star goes wild and varies chaotically (Figure 13.9, bottom).

Beyond a second critical dynamo number, $D2$, we see the heights of the peaks modulated in long cycles, as in Figure 13.9c. Finally, when we reach the next critical rotation rate, at D3, the star goes wild and varies chaotically (Figure 13.9d).

Chapter 14

THE SUN AND CLIMATE

14.1 Carbon 14 Dating

As everyone knows, trees grow by producing cellulose from the carbon dioxide in the atmosphere. Carbon has several forms called isotopes that have identical chemical properties and differ only in their atomic weight. Carbon atoms with a weight of twelve (^{12}C) are stable, those with a weight of fourteen (^{14}C) are radioactive and decay at a steady rate over a period of about 30,000 years.

In 1947, Willard Libby, a chemist at the University of Chicago, was the first to date samples of wood using the decay of ^{14}C. The ratio of ^{14}C to ^{12}C in a sample measures the time since the ^{14}C was absorbed in the wood, or in other words, its age. Radiocarbon dating became an essential tool for archaeologists and Libby received the Nobel Prize for his work in 1960.

14.2 How Carbon 14 Forms

Our atmosphere is constantly bombarded by galactic cosmic rays. When these energetic charged particles strike the gases in the atmosphere, they produce showers of secondary nuclei, such as radioactive isotopes of carbon (^{14}C), oxygen (^{18}O), beryllium (^{10}Be), and other light elements.

As you may recall, solar wind flows continuously around the earth. When the sun is active, and sunspots are abundant, the strong magnetic field in the wind shields the earth from cosmic rays and carbon 14 production is then severely reduced. During solar minima, the wind's field is weaker, the shielding effect is reduced, and the cosmic rays can arrive in full strength. So the ^{14}C wiggles indicate the level of solar activity.

14.3 GISP2 Analysis

In Figure 14.6 we see how they compared their measured ^{18}O fluctuations to sunspot numbers, ^{10}Be concentrations, and the production rate Q of ^{14}C, over the time span from 1600 to the present.

The Maunder Minimum stands out clearly in the sunspot record from 1650 to 1700. What can one see in the other records that corresponds to this Minimum? The amplitudes of the ^{18}O wiggles are smaller (because of colder temperatures) and those of the ^{10}Be wiggles are larger (because of easier cosmic ray entry), just as one might expect. The same is true for the so-called Dalton Minimum, around 1800. But, surprisingly, the ^{14}C production rate Q (the *difference* in the amounts of ^{14}C in tree rings in successive years) shows practically no sign of a Maunder Minimum.

The best test of an eleven-year cycle is how well the wiggles in the different records match up. In Figure 14.6, Stuiver overlaid each record on the sunspot record. The match of peak and valley is quite impressive, indicating a strong eleven-year cycle in the amount of ^{18}O in the ice. Indeed, statistical tests showed a very strong correlation between sunspot numbers and Greenland's temperature. But to get this nice result, the ^{10}Be and Q records had to be shifted forward in time by two years, and the ^{18}O record *retarded* by two years. Stuiver and company explained these shifts in terms of storage times of the isotopes in the atmosphere, but their arguments seem somewhat contrived.

14.4 Land Temperatures and the Solar Cycle

To extend their connection further back in time, the Danes first used the quantity of sea ice around Iceland as a proxy for cold temperatures in the North Atlantic. They showed that cycles longer than eleven years correspond to colder temperatures, and shorter cycles to warmer temperatures, from 1750 to 1990.

Consistent observations of sunspots were pretty "spotty" before 1750. Therefore to extend their connection even further back in time the Danes used observations of auroras, going back to 1500, to determine the minima of sunspot activity, and from these, the lengths of the solar cycles. Then they compared them with another scientist's reconstruction of Northern Hemisphere temperatures, reaching back to the mid-1500s. Sure enough, their correlation between cycle length and air temperature held up nicely.

14.5 The Importance of Ozone

You may recall that ozone is a molecule, composed of three oxygen atoms, that absorbs solar ultraviolet light. (The destruction of the ozone layer over the Antarctic by CFCs has been a recent worry of environmentalists.) In a complicated chain of events, changes in the ozone layer affect stratospheric winds, which in turn affect the wind and temperature distributions in the lower atmosphere (the "troposphere"). The question that has to be answered is whether the eleven-year variation of ultraviolet light (amounting to 10–20%), and of the ozone layer, is sufficient to drive changes in the climate.

Drew Shindell and his colleagues at the NASA Goddard Institute of Space Studies took a hard look at this question. They used a powerful computer model of the earth's atmosphere, which includes a detailed description of the photo-chemistry of ozone, and the formation of so-called planetary waves in the lower atmosphere. Their calculations extended over a simulated time of twenty years, including a solar maximum and minimum. They showed how changes in the circulation of the stratosphere, driven by changes in solar ultraviolet, penetrate into the troposphere. The model reproduced several known eleven-year oscillations of the atmosphere, including the changes of the temperature distribution in the troposphere discovered by Labitzke and van Loon.

GLOSSARY

accuracy—The degree to which a measurement approaches its true value

black body—A hypothetical body that absorbs all wavelengths of light equally well

convection—A process of transporting heat by the bodily movement of hot cells of gas or liquid

convection zone—A region in the sun where heat is transported mainly by convection

Coriolis effect—The apparent lateral drift of an object that moves relative to an observer who is located on a rotating body

coronagraph—A special telescope that produces artificial eclipses of the sun

diagnostic diagram—A display of solar acoustic modes, arranged by horizontal wavelength and frequency

Doppler effect—The shift in frequency of a source of light or sound as it passes a fixed observer

flux, magnetic—The quantity of magnetic field lines within a fixed area

flux, radiative—The amount of radiative energy passing through a fixed area

force-free magnetic field—A field in which any electric currents flow parallel to the field lines, and generate no net Lorentz forces

gauss—A unit of magnetic field strength. The earth's field is about half a gauss at the magnetic poles.

granules—A pattern of convective cells, each about 1,400 km in diameter, that lies near the solar surface

kelvin—A unit of temperature, equal to one degree centigrade. The zero point is −273 degrees C.

luminosity—The total output of energy of a source

magnetograph, magnetometer—A device for displaying the strength and polarity of magnetic fields

maxwell—A unit of magnetic flux

microhertz—A frequency of one-millionth of a cycle per second

mode—A distinct pattern of vibration; a harmonic or tone

momentum—In mechanics, the product of mass and velocity. The tendency of a body to maintain its current motion.

nanohertz—A frequency of one-billionth of a cycle per second

nanometer (nm)—A billionth of a meter

node—A point in a vibrating system that remains stationary

photosphere—The radiant surface of the sun

plasma—A gas consisting of ions, electrons, and neutral atoms, in which electrical forces are dominant

polarization—The preferred orientation in space of the electric vibrations in a light beam

potential magnetic field—A field in which no electric currents flow. A bar magnet contains the atomic currents that produce the potential field outside the bar.

precision—The lack of uncertainty in a measurement

spectral line—A characteristic emission of an atom, at a precise wavelength, that corresponds to a jump of an electron between two energy levels in the atom

spectrograph—A device for analyzing light by producing a spectrum

spectroheliograph—A device for imaging the sun in the light within a narrow band of wavelengths

spectroscopic polarimetry—The measurement of the polarization of the light within a spectral line

spectrum—A display of the intensity of radiation, sorted according to wavelength

standing wave—A wave system that produces oscillations at fixed points, with no propagation of phase

streamer—A fan of coronal plasma with a characteristic petal-like shape

supergranulation—A pattern of convective cells, each about 30,000 km in diameter, that lies near the solar surface

Zeeman effect—The characteristic splitting of an atom's spectral lines in a magnetic field

INDEX

Page numbers in italics refer to figures.